Hume's Science of Human Nature

"Hume's Science of Human Nature offers, and capably defends, a deeply original way of understanding Hume's theoretical framework. The interpretation renders Hume's naturalistic project a much more sophisticated endeavour than typically thought. It is an excellent and welcome addition to the literature."

—**Hsueh Qu,** *National University of Singapore*

Hume's Science of Human Nature is an investigation of the philosophical commitments underlying Hume's methodology in pursuing what he calls 'the science of human nature'. It argues that Hume understands scientific explanation as aiming at explaining the inductively-established universal regularities discovered in experience via an appeal to the nature of the substance underlying manifest phenomena. For years, scholars have taken Hume to employ a deliberately shallow and demonstrably untenable notion of scientific explanation. By contrast, *Hume's Science of Human Nature* sets out to update our understanding of Hume's methodology by using a more sophisticated picture of science as a model.

David Landy is Associate Professor of Philosophy at San Francisco State University. He is the author of *Kant's Inferentialism: The Case Against Hume* (Routledge, 2015).

Routledge Studies in Eighteenth-Century Philosophy

For a full list of titles in this series, please visit www.routledge.com

Books in this series:

7 **Aesthetics and Morals in the Philosophy of David Hume**
Timothy M. Costelloe

8 **Hume's Difficulty**
Time and Identity in the Treatise
Donald L.M. Baxter

9 **Kant and the Cultivation of Virtue**
Chris W. Surprenant

10 **The Post-Critical Kant**
Understanding the Critical Philosophy through the Opus postumum
Bryan Wesley Hall

11 **Kant's Inferentialism**
The Case Against Hume
David Landy

12 **Rousseau's Ethics of Truth**
A Sublime Science of Simple Souls
Jason Neidleman

13 **Kant and the Scottish Enlightenment**
Edited by Elizabeth Robinson and Chris W. Surprenant

14 **Kant and the Reorientation of Aesthetics**
Finding the World
Joseph J. Tinguely

15 **Hume's Science of Nature**
Scientific Realism, Reason, and Substantial Explanation
David Landy

Hume's Science of Human Nature

Scientific Realism, Reason, and Substantial Explanation

David Landy

Routledge
Taylor & Francis Group
LONDON AND NEW YORK

First published 2018 by Routledge

2 Park Square, Milton Park, Abingdon, Oxfordshire OX14 4RN
52 Vanderbilt Avenue, New York, NY 10017

Routledge is an imprint of the Taylor & Francis Group, an informa business

First issued in paperback 2019

Library of Congress Cataloging-in-Publication Data
A catalog record for this book has been requested

ISBN: 978-1-138-50313-7 (hbk)
ISBN: 978-0-367-89171-8 (pbk)

Typeset in Sabon
by Apex CoVantage, LLC

The form of this book is dedicated to Jay Rosenberg, who taught and encouraged me to do philosophy like a surprisingly responsible drunken reveler. Its content is dedicated to Alan Nelson, who treated me as a colleague when I was his student, resulting in what follows. Finally, the entire book is dedicated to Margo, who makes everything that I do not only possible, but quite pleasant.

Contents

Acknowledgments

Portions of the following chapters have been previously printed as follows. My thanks to these publishers for allowing me to reprint this material here.

Chapter 1

"Recent Scholarship on Hume's Theory of Mental Representation," *European Journal of Philosophy*, forthcoming.

"A Puzzle about Hume's Theory of General Representation," *Journal of the History of Philosophy*, 54, 2 (April 2016): 257–82.

"Hume's Theory of Mental Representation," *Hume Studies*, 38, 1 (April 2012): 23–54.

Chapter 2

"Is Hume an Inductivist?" *Hume Studies*, 41, 2 (November 2015): 231–61.

Chapter 3

"Sellars and Hume on the Language of Theories," In *Sellars and the History of Modern Philosophy*. Eds. Antonio Nunziante and Luca Corti. New York: Routledge, forthcoming.

Portions of this manuscript were written with the generous support of the San Francisco State University Sabbatical Award. The book benefitted greatly from the feedback I received on material related to it presented in the following venues: the West Coast Hume Workshop, the 41st International Hume Society Conference, and department colloquia at the University of California at Santa Cruz, Occidental College, the University of Oklahoma, and Wayne State University. Comments on drafts from Jonathan Cottrell and Hsueh Qu were also very helpful, stimulating, and insightful. As always, Margo Landy provided an enormous amount and variety of support. Any faults that remain after the contributions of so many helpful readers are of course all my own.

Notes on Hume's Texts

For citations from Hume's *A Treatise of Human Nature* and *Enquiry concerning Human Understanding* I employ the standard convention of citing the book, chapter, section, and paragraph number from the Clarendon edition, followed by the page number from the Selby-Bigge/Nidditch edition. I also employ the following abbreviations.

Abs	*An Abstract of a Book Lately Published.*
T	*A Treatise of Human Nature.*
EHU	*Enquiry concerning Human Understanding.*
D	*Dialogues Concerning Natural Religion*
El	"Of Eloquence" in Miller, *Essays: Moral, Political, and Literary.*
H	*History of England.*
Is	"Of The Immortality of the Soul" in Miller, *Essays: Moral, Political, and Literary.*
N	*The Natural History of Religion* in Beauchamp, *A Dissertation on the Passions* and *The Natural History of Religion.*
Sc	"The Sceptic" in Miller, *Essays: Moral, Political, and Literary.*

Introduction

This book is an attempt to interpret Hume's approach to what he calls the science of human nature by way of his understanding of scientific explanation. Hume repeatedly frames the aim of his science to be to produce explanations, and thus understanding how he thinks of that goal ought to be a good way to give shape to how he goes about pursuing it. Additionally, there has been a fair amount of recent scholarship on Hume's account of scientific explanation, so the ground is fertile for discussion, but I believe that at least some of that scholarship has significantly missed its mark. By way of illustration consider two views that reside at opposite ends of the interpretive spectrum.[1]

On the one hand, at the turn of the previous century the so-called Logical Positivists or Logical Empiricists appropriated Hume as their historical forefather, casting him as endorsing something very much like their own account of scientific explanation, the Deductive-Nomological account (DN).[2] According to DN, the *explicandum* of a scientific explanation is an observed particular, and its *explanans* is an inductively-established universal generalization. For example, we observe the position of Mercury in the night sky and explain this position by deducing it from Kepler's laws of planetary motion and the previously observed position of Mercury. Scientific explanation and progress takes the form of discovering propositions of increasing generality that subsume an increasing variety of particular phenomena under them.

That picture of scientific explanation has long been recognized as inadequate, and yet the belief that Hume holds a view very similar to it has proven remarkably and unfortunately durable. I believe that something like it is an implicit presupposition of many current approaches to Hume, and a version of this interpretive line has recently been defended by Graciela De Pierris by way of her defense of Hume as adopting Newton's so-called inductivist account of scientific methodology. Before turning to the specific inadequacies of such an approach as an interpretation of *Hume*, it will be helpful to have an alternative account of scientific explanation itself at hand, and throughout this study I will use the influential account of scientific explanation delineated and defended by Wilfrid Sellars as a touchstone.

More on the motivation and details of that choice in a moment. For now, consider that as Sellars understands it, DN makes two important mistakes regarding scientific explanation. The first is misidentifying both its *explicandum* and its *explanans*. As Sellars sees it, the *explicandum* of a scientific theory is not observed particulars, but rather the fact that the particulars are subject to the empirical generalities that they are. So, in the example above, it is not the position of Mercury alone that demands explanation, but rather the regular motions of the planets, or Kepler's law itself. Or to take an example from Hume, as we will see, when he leverages the simple-complex distinction in explicating the Copy Principle—that all *simple* ideas are copied from some corresponding impression—it is not the occurrence of this or that perception that requires explanation, but rather the great novelty and systematic connectedness of human thought given the *prima facie* plausibility of the thesis that all ideas are derived from impressions. Furthermore, as Sellars sees it, the *explanans* of such phenomena makes essential appeal to the nature of the particulars at hand:

> theories about observable things do not "explain" empirical laws in the manner described, they explain empirical laws by explaining why observable things obey to the extent that they do, these empirical laws; that is, they explain why individual objects of various kinds and in various circumstances in the observation framework behave in those ways in which it has been inductively established that they do behave. Roughly, it is because a gas is—in some sense of 'is'—a cloud of molecules which are behaving in certain theoretically defined ways, that it obeys the empirical Boyle-Charles law.[3]

Along these lines, the ultimate explanation for the motions of the planets comes via an explanation of the law of universal gravitation as the manifestation of the curvature of spacetime. Hume's simple-complex distinction explains the novelty and systematicity of human thought by appealing to the nature of perceptions themselves: that our complex phenomenology is *composed of* simple perceptions united according to certain associative tendencies. In both cases, the inductively-established universal regularities are explained via an appeal to the nature of the object at hand to "explain the why observable things obey to the extent that they do, these empirical laws."

Misunderstanding the proper *explicandum* and *explanans* of a scientific theory in turn leads to the second important mistake that Sellars finds in DN: it overlooks the methodological importance of explanation's dependence on *perceptible models*.

> the fundamental assumptions of a theory are usually developed [. . .] by attempting to find a *model*, i.e. to describe a domain of familiar objects behaving in familiar ways such that we can see how the phenomena

> to be explained would arise if they consisted of this sort of thing. The essential thing about a model is that it is accompanied, so to speak, by a commentary which *qualifies* or *limits* [. . .] the analogy between the familiar objects and the entities which are being introduced by the theory.[4]

On Sellars's understanding of theoretical explanation, then, one explains the fact that certain particulars are subject to certain empirical generalizations via an appeal to the fact that these particulars have—as we will see Hume put it—certain qualities, natures, or powers. These qualities of the particulars, in turn, are represented via perceptible models, which are specified to resemble and differ from the posited entities in determinate ways. To switch examples for a moment, consider a familiar case of such modeling from the sciences: the Bohr model of the atom. Taking the solar system as its perceptible model, Bohr proposes that we understand the structure internal to an atom as analogously consisting of a large nucleus being orbited by smaller electrons, and also specifies determinate differences between the theoretical entity posited and its perceptible model: the former is orders of magnitude smaller, it is held together by electrostatic forces instead of gravity, the electrons' orbits are circular rather than elliptical, etc. As foreign as it might seem to contemporary scholars still in the grips of the Positivist picture of Hume's understanding of scientific theorizing, what I hope to show here is that something like Sellars's understanding of scientific explanation is much closer to Hume's own approach to the science of human nature than has previously been understood.

Of course, there are many positions possible in this space other than these two. For example, if one accepts that it is empirical regularities that are the *explicanda* of scientific theories, and that it is the nature, powers, and essences underlying manifest phenomena that are their *explanans*, but *rejects* that the latter are represented by perceptible models, or any representation with descriptive content at all, then one more or less arrives at the cluster of views commonly referred to as those of the 'New Hume',[5] which occupy a place at the opposite end of the interpretive spectrum from those of the Logical Positivists.

For example, as Galen Strawson interprets him, Hume takes the proper *explicandum* of scientific explanation to be the inductively-established universal regularities discovered in experience, and the proper *explanans* to be the nature, power, or qualities of the substance underlying such manifest phenomena. As will become clear, I believe that Strawson has that much right. Where I believe New Humeans go wrong, though, is in thinking that we cannot form any idea with descriptive content of such substances or causal powers, but can at most refer to such explanatory entities merely as that which explains these regularities.[6] In addition to what I take to be the interpretive problems with such a view there is also the more purely philosophical problem that this is not an account of scientific *explanation* at all. Even if we could successfully refer to that which explains the observed phenomena using no description at all (which I do not believe that we can), what one would have upon doing so is not an explanation of those phenomena, but rather only a promissory note that such an explanation exists. That is, the end product of such reference is not an explanation itself, but only a reference to some explanation we know not what.

Regardless of the merits or demerits of such a view, I do not believe that it is Hume's. Instead, I will argue over the course of this book that Hume's view occupies a middle ground on the spectrum between De Pierris's "inductivist" interpretation of him and that of the New Humeans. For Hume, scientific explanation consists of more than mere inductive generalization because he holds that such generalizations are the proper *explicanda* of scientific explanation, not its *explanans*. An explanation of such regularities will make appeal to the nature and powers of the substance underlying them. Hume does not hold, however, that such an explanation is achieved merely by making bare reference to such a substance. Rather, that which explains these universal regularities, to be genuinely explanatory, must describe this substance in a way that shows *how* its nature or powers account for the observed phenomena. Such a description, in turn, can only be achieved by forming a representation with genuinely descriptive content, and for Hume this means deriving that content from experience itself in some suitable sense.

Scholars from both ends of this interpretive spectrum will share common ground in objecting to this suggestion: they will both cite the close tie that Hume draws between representation and experience that appears to preclude representing such theoretical entities. One reason to think that Hume would hold that explanation consists of subsuming particulars under inductively-established universal generalizations is that he takes the content of our ideas to be derived in a very *direct* way from our impressions. If all we can think is what we can experience, the reasoning goes, then the most that we can do by way of explaining observed phenomena is appeal to the regularities observed therein and inferred therefrom. If all the representational resources that we have derive from observation, then we cannot represent anything that is unobservable. The New Humean concurs with this assessment, but adds that while we cannot *represent* anything unobservable, we can still *refer* to such things, although not by employing any idea with descriptive content.

Again, I will argue that there is a middle ground between these two positions that both accords with Hume's texts and makes his understanding of scientific explanation and theoretical representation more plausible. My proposal is that such representations are formed by taking experience itself as a perceptible model, and specifying the determinate ways that the theoretical entity posited both differs from and resembles this model. Consider again our toy example: modeling the structure of the atom on that of the solar system. What we do in such a case is take the observed structure of the solar system as a model and form a theoretical representation of the atom by specifying determinate ways that this posited explanatory entity both differs from and resembles that model. Both consist of a large central body being orbited by smaller particles, but whereas the solar system is held together by gravitational forces, electrons are kept in place by electrostatic ones, etc. This form of explanation both moves beyond merely describing the universal regularities discovered in experience by appealing to the nature and powers of the substance underlying the manifest phenomena, but does so not via a bare reference to that which explains such regularities, but rather by forming a determinate representation of that substance with descriptive content via the use of such a perceptible model.

Of course, there is a great deal for Hume to resist in this particular example, and there are a great many details to be filled in if such a picture of scientific explanation is to be made properly Humean, but I will argue that something like this understanding of scientific explanation, at its most general level, is precisely what underlies Hume's pursuit of the science of human nature. In fact, I will argue that we *must* take Hume to have some such understanding of the explanatory and representative resources of science if we are to make sense of the most fundamental parts of his philosophical system.

Realism

Of course, the spectrum delineated by placing De Pierris and Strawson at either extreme is not the only one on which my interpretation is located. Consider, for example, that one way to describe the sketch of a view that I have just drawn is as a kind of scientific *realism*—what science purports to represent is the *reality* underlying manifest appearances—and 'realism' is itself a term with its own history in Hume scholarship. Kail's *Projection and Realism in Hume's Philosophy* is helpful here, first, because Kail rightfully notes that one is not a realist or anti-realist *simpliciter*, but rather always a realist or anti-realist about some specific domain. For example, one can be a realist or anti-realist about necessary connection, or the external world, or the self, or primary qualities, secondary qualities, or God, etc. And one can be a realist or anti-realist about any of these domains with respect to various sub-domains as well. For example, one can be a realist or anti-realist about the external world *as it is represented by* the vulgar, or by the false philosopher, or by the scientist of human nature. What I will argue, is that Hume is a realist about the objects posited by the science of human nature and the differences in kind between these, e.g., impressions and ideas, simples and complexes, the faculties of memory, imagination, and reason. Such objects and distinctions cannot be directly observed, but are theoretical posits made and accepted by the scientist of human nature because of their explanatory force.

Additionally, though, because one of the conclusions that the scientist of human nature reaches is that we can never represent anything that is not a perception (because we cannot specify *determinate differences* between perceptions and non-perceptions), Hume is also a fairly radical kind of *anti*-realist about certain other domains. Specifically, I will argue that while Hume understands the science of human nature to use language to picture correctly the qualities, nature, and powers underlying the human mind, he understands the language of the vulgar and of the false philosophy to be failed attempts to do the same. Thus, the pictures that they each attempt to form of the external world, necessary connection, and the self must be abandoned in favor of that of the scientist of human nature, and that picture represents no such things.

Understanding this form of anti-realism is a second place that Kail is particularly helpful. In addition to delineating various ways in which one might be a realist, he is also careful to parse various senses of its converse, anti-realism. Regarding the latter, he lists four common forms of anti-realism often encountered in the literature on Hume. (His focus is on anti-realism with respect to the external world and God.)

(1) Reductionism rejects the core content for the respective belief and replaces it with analysis of such commitments couched in terms of allegedly more tractable materials.
(2) Non-cognitivism holds that statements regarding God or the external world express statements of mind, the function of which is other than representing the world to be thus and so.
(3) Rejection [. . .] is the view that such beliefs lack justification and without it we should not believe.
(4) [D]eep incoherence. There are not coherent thoughts to be had concerning the external world or God.[7]

By this reckoning, then, my interpretation of Hume's view of the external world, necessary connection, and the self counts as a form of deep incoherence view. I do not hold that such talk is *really just* talk of contingently connected perceptions because I hold that the language of the vulgar and that of the false philosophy at least purport to represent something over and above these perceptions, and even if they fail in that purport, the representations that they do produce are not *reducible* to a mere description of observable phenomena. Rather the languages of the vulgar and the false philosophy are *incompatible* with such a description, to their detriment, and so must be abandoned in favor of the language of the scientist of human nature, which also represents more than what is phenomenologically available to us.

My interpretation is also neither a form of non-cognitivism nor a form of error theory. I take Hume to understand the languages of the vulgar and false philosophies as forming genuine representations of the world that are not the mere expressions of their states of mind, although it is their states of mind that serve as the *models* of their theoretical representations. Contrary to Blackburn, for example, I do not understand the vulgar's or false philosopher's talk of necessary connection as merely expressing habits of inference.

Again, these languages at least purport to represent necessity as a property of "the real world." That said, their error in doing so also amounts to more than merely lacking the proper justification. It is not that, for example, the vulgar and false philosophers err insofar as their beliefs are not supported by reason, but by habit alone. Hume's rejection of the explanations offered by the vulgar and false philosophers is not based on their being *groundless*, but rather on their being (depending on the case) *false*, *contradictory*, or without *determinate content*.[8] Again while the details of Hume's rejections of these attempts at explanation differ from case to case, in none of these cases it is that the picture at hand merely lacks the proper justification. In some, parts of the views are straightforwardly refutable by empirical evidence, but for most the problem runs deeper. Almost all fail because while they *attempt* to picture reality using perceptions as models, they are unable to specify *determinate* similarities and differences between this model and their theoretical posits, and thus their representations fail to have the robust content that they would need to successfully do so. Thus, Hume concludes that there is indeed a deep incoherence to the view at hand because the representations that the vulgar and false philosophy attempt to use lack the proper descriptive content, they end up not representing anything determinate at all.

To summarize, then, I will argue that Hume is a *realist* about that which the science of human nature represents as the nature, powers, essence, or substance of the human mind, but that he also holds that there is a deep incoherence to the explanations offered by the vulgar and false philosophers, and so he is an anti-realist about the external world, necessary connection, and the self.

Skepticism and Naturalism

Having outlined where my interpretation will come down with respect to scientific explanation and the various senses of realism and anti-realism, it is also worth addressing one final cluster of related issues: those pertaining to Hume's skepticism, his naturalism, and his understanding of the role of reason in our epistemic endeavors. While many scholars have grappled with the tensions in Hume's writing surrounding these issues (e.g., Loeb, *Stability and Justification*, Noonan, *Routledge Philosophy Guidebook to Hume on Knowledge*, and Owen, *Hume's Reason*), it will be helpful to have a single focal point with which to compare the current study, and to that end I will use Garrett's *Cognition and Commitment*.

As Garrett reads Hume, his project consists primarily in an empirical investigation into our *cognitive psychology*, although this investigation also yields certain normative epistemological conclusions as well.[9] With respect to the former, consider Garrett's understanding of Hume's skepticism about induction.

> Hume should be interpreted quite literally, as making a specific claim, within cognitive psychology, about the relation between our tendency

> to make inductive inferences and our inferential/argumentative faculty: he is arguing that we do not adopt induction on the basis of recognizing an *argument* for its reliability, for the utterly sufficient reason that there *is* no argument ("reasoning" or "process of the understanding") that could have this effect.[10]

Hume is a skeptic about induction not because he thinks that inductive inferences are unjustified—he does not—but rather because he holds that it is not any operation of our inferential faculty that itself justifies them. What does justify them is what Garrett calls the Title Principle: "where reason is lively and mixes itself with some propensity, it ought to be assented to" (T 1.4.7.11; SBN 270).[11] Thus, Garrett balances Hume's skepticism against his naturalism and relates them both to his understanding of reason. Hume is pursuing a project in cognitive psychology that includes an investigation into the cognitive mechanisms that we call "reason". That investigation yields the conclusion that there is no argument (activity of the inferential faculty) the conclusion of which is that that faculty itself is truth conducive. Nonetheless, Hume does not reject the use of that faculty—such a rejection certainly would not be supported by the conclusion of his investigation alone—but instead appeals to the mechanisms that govern our actual use of that faculty.

To begin, I will say that I agree with much of Garrett's picture. I too hold that Hume is engaged in a project concerning cognitive psychology and that one form that his skepticism takes is a denial of reason's ability to justify its own use. There are a number of points, however, where my interpretation differs from Garrett's. First and foremost, whereas Garrett takes our faculty of reason to comprise only our deductive and inductive inferential activities, I understand Hume as also countenancing a form of inference to the best explanation. For example, as I have indicated, I will argue that Hume accepts the distinction between simple and complex perceptions not because we can *observe* the difference between these (we never perceive a simple perception as such), but rather because such a distinction *explains* the enormous variety and systematic connectedness of human thought in the face of the relative paucity of human experience. That being so, I understand Hume's pursuit of the science of human nature as having a wider scope and authority than does Garrett, specifically, it includes not only a description and systematization of our phenomenology, but also speculation concerning the *explanation* of the regularities observed in that phenomenology that makes appeal to the nature, powers, essence, or substance underlying it, the belief in which is justified on the basis of its explanatory force. That expansion of Hume's naturalism, in turn, also affects the scope of his skepticism.

Again, while I agree with Garrett that Hume's skepticism is limited to his thesis that deductive and inductive reason cannot justify their own uses, I also hold the aforementioned related thesis that Hume is a deep incoherence anti-realist with respect to several of the *explanans* employed by the vulgar and false philosophers. One way that skepticism with respect to a certain domain is understood is as consisting of the thesis that some set of propositions concerning that domain cannot be *known*. To get to Garrett's position from there, one need only first notice that whatever else is required for knowledge, having a justified belief appears to be one of its necessary conditions. Focusing on the nature of the justifications available (or not) for our beliefs concerning the reliability of inductive and deductive inference then leads Garrett to Hume's mitigated skepticism about these. All of that, however, supposes that there is some proposition or belief with the kind of content that can or cannot be known. As I will argue, though, when it comes to the explanations offered by the vulgar and false philosophers, part of what I take to follow from the requirements that Hume puts on theoretical explanations, is that these explanations are, in fact, without representational content. Because they do not specify determinate resemblances and differences between their theoretical posits and the perceptions on which these are modeled, the purported representations fail to represent anything determinate at all. Thus, I take Hume's naturalism to lead not only to the mitigated skepticism with respect to reason that Garrett describes, but also to a much more radical rejection of certain other theses. On the other hand, since I also understand Hume as holding that the science of human nature as he pursues it is mostly *successful*, I also read him as endorsing the deep ontological implications of that science, for example, to real distinctions of kind between simple and complex perceptions, impressions and ideas, and our faculties of memory, imagination, and reason. So, Hume is, in turn, a naturalist, a mitigated skeptic, a deep incoherence anti-realist, and a scientific realist.

The Place of Sellars in This Interpretation

As I mentioned earlier, this rather unorthodox interpretation of Hume is itself modeled on the understanding of scientific explanation defended by Wilfrid Sellars, and as such a word is in order regarding the motivation for that decision and its consequences. To begin, I should be up front about the fact that I have spent a good number of years learning the ins and outs of Sellars's intricate, controversial, and powerful philosophical system, beginning with my mentorship under Jay Rosenberg, himself Sellars's first graduate student at the University of Pittsburgh and lifelong Sellars expounder. So, I am almost certainly predisposed towards thinking that a Sellarsian reading of a philosophical text is the most charitable one![12] Additionally, though, Sellars's understanding of scientific explanation serves as a useful tool for understanding Hume for at least two independent reasons. First, Sellars, like Hume, holds a *picture theory* of representation that is essential to his account of scientific explanation and progress. Hume's and Sellars's pictures have different constituents and different structures—for Hume the elements are simple ideas and the structure their associations, for Sellars natural linguistic objects and their inferential relations—but the fact that both are picture theories itself makes them each subject to similar constraints and allowances. Thus, using Sellars's understanding of scientific explanation, which is formulated within the context of his picture theory of representation, as a model for Hume's carries with it the distinct advantage of having many of these features already built into it. Relying on an alternative understanding of scientific explanation that was not built on such a theory would require building the connections between these two aspects of Hume's philosophical system from scratch with no guarantee that this would be possible.

Second, part of my interpretive project here is to reject the longstanding, if often implicit, link between Hume's understanding of the science of human nature and that of the Logical Positivists, while also fending off the opposing pressures posed by the New Humeans. Sellars fits nicely into this logical space because part of *his philosophical* project is to reject the account of scientific explanation defended by Logical Positivists (and attributed by them to Hume), while at the same time resisting the (as of then still nascent) opposing pressures posed by the Kripke/Putnam accounts of scientific realism via so-called "direct reference" and "rigid designation." That my interpretive project is located in this dialectical, or logico-structural, space analogous to the one that Sellars's philosophical project is, means that again I expect to find resources in Sellars's work that speak directly to the positions that I find Hume taking up. Those expectations have not been disappointed, although I again must admit that this could be the result of biased rather than clear thinking. The only defense against that charge, of course, is to let the interpretation stand on its own, and to see how it does.

I will turn to that task in a moment, but before doing that it will be worth briefly cataloging some of the most significant similarities and differences between my model, Sellars's understanding of scientific explanation, and my interpretive object, Hume's.

- I have already mentioned that I take Hume and Sellars to share a conception of scientific explanation according to which inductively-established universal regularities of experience that are its proper *explicanda*, and that its proper *explanans* will be the nature, powers, or essence of the substance underlying these.
- I have also already mentioned that I take Hume and Sellars to share the thesis that such theoretical-explanatory posits must be represented via a perceptible model by specifying determinate ways in which the posit both resembles and differs from that model.
- Such explanations function to explain the manifest appearances by appealing to the reality that underlies them, and Hume and Sellars also both take such explanations, when successful to carry with them the most fundamental form of ontological commitment. In Sellars this takes

the form of his famous *scientia mensura*: "In the dimension of describing and explaining the world, science is the measure of all things, of what is that it is, and of what is not that it is not."[13] I understand Hume as holding a version of this thesis as well, although with an important caveat. Because the science of human nature is the most fundamental of the sciences, it is that science in particular that is the ultimate arbiter of what does and does not exist. (As it will turn out, once those conclusions are reached, much of what the other sciences have postulated will have to be rejected or reconceived.)

- Relatedly, I also take Hume and Sellars to share a view of what it looks like to assume such ontological commitments. Sellars holds that with respect to the language of science,

> the Scientific Realist need only argue [. . .] that in principle this language could *replace* the common-sense framework in *all* its roles, with the result that the idea that scientific theory enables a more adequate picturing of the world could be taken at its face value.[14]

Correspondingly, I will argue that in Hume's system our commitment to the ontology of science manifests itself in our adopting the language of the scientist of human nature (as opposed to that of the vulgar or the false philosophy) and thereby endorsing the distinctions drawn by the scientist as corresponding to some real resemblance relations between different kinds of mental particulars and faculties.

Those are the most salient and important similarities between Sellars's conception of scientific explanation and Hume's as I read them, although less significant similarities certainly arise as this study unfolds, and the differences between the two will also assume some importance. Those points, however, concern matters of detail that can only be addressed during the study itself, and so it is to a summary of that study that I will turn in a moment before one final comment on the scope and nature of what is to follow.

For the most part, my focus here will be almost entirely on Book I of the *Treatise*. This is for a number of reasons. To begin, I hold that, despite Hume's desire that his *Enquiries*, as opposed to the *Treatise*, be taken as the canonical expression of his philosophical system, the *Treatise* nonetheless remains a worthy object of scholarly interest. I do, however, also accept the implication of Hume's preference: that the *Enquiries* and *Treatise* are sufficiently *different* that the claims and arguments made in one of these cannot be assumed to be the same as the claims and arguments made in another. Thus, while it is certainly true that Hume has a great deal to say about his understanding of his pursuit of the science of human nature in the *Enquiry*, whether or not what he says there reflects the understanding he expresses in the *Treatise* cannot be taken for granted. Rather, the only way to answer that question is to conduct *independent* scholarship on the *Treatise* and

Enquiry, respectively, and then compare the results. The current study can be understood as an attempt at the first half of that process.

As indicated, though, I restrict my scope here not just to the *Treatise*, but more narrowly to Book I. Here the restriction is less for exegetical reasons than architectonic ones. As I understand it, Book I lays the foundation for Books II and III, and as such it should be possible to understand Book I prior to understanding what follows. Given all of the intricacies of the former task, the work of expanding the conclusions reached here to the arguments and conclusions presented there would require more care and attention than a reasonable-length study allows. That said, because the conclusions of Book I become the premises of Books II and III, there are moments where looking to Books II and III to see what further implications Hume understands these conclusions to carry is of some use, and I take advantage of those moments as they arrive, but again only instrumentally, as they are of help in understanding Book I. I look forward to expanding the scope of the current study into these areas at some time in the future, but for now, onto the matter at hand.

Chapter Outlines

In Chapter 1, I argue that both the impression-idea and simple-complex distinctions only do the explanatory work that Hume needs them to do if we understand them as implicitly deploying theoretical-explanatory representations. This argument begins with a puzzle concerning Hume's theory of general representation. After reviewing the recent scholarship on Hume's theory of mental representation,[15] I conclude that for Hume the content of a general representation is that the objects of the ideas that compose the revival set of a general term all resemble one another.[16] I defend what I call the Representational Copy Principle, the two-part thesis that all simple ideas represent that of which they are copies, and that all complex ideas represent the objects of their component simple ideas as being arranged in the way that the complex idea is itself arranged. To capture this latter part of the thesis, I employ the following schema:[17]

'x'R'y' represents xRy.

To represent x as related to y in relation R, one forms a complex idea consisting of an idea of x related to an idea of y in relation R. For example, to represent x as next to y, one forms a complex idea consisting of an idea of x next to an idea of y. That is, we represent spatial relations by forming mental pictures of them. To represent x as preceding y, one forms a complex idea consisting of an idea of x preceding an idea of y—i.e., we represent temporal relations by forming mental movies of them, etc. What follows from this is that to represent two items as resembling each other, one must form an idea of each item, which ideas are themselves related by a distinctive resemblance relation. As it is this relation that Hume argues holds between members of the revival set of a general term, it follows that what a general representation represents is that the members of its revival set resemble one another.

Consider, though, that because our phenomenology is always and everywhere *complex*, the ideas that compose the revival set of the term, 'simple idea' will all be complex ideas. If, however, the distinction between simple ideas and complex ideas is to help explain the dependence of the variety of possible human thought on the relative paucity of human experience, then "simple idea" cannot just mean those-complex-ideas-associated-with-the-term-"simple-idea," but must rather pick out the *simple components* of these ideas, which I construe as theoretical posits. Given that we never encounter simple ideas *as such* in experience, there is a puzzle about just how this reference can be secured. Thus, Hume needs an account of theoretical representation that employs resources beyond those afforded merely by his theory of general representation. Furthermore, what such resources will secure for Hume is an explanation that makes essential appeal to the "nature, powers, or essence" of the substance underlying experience. That is, I conclude that Hume is a kind of *scientific realist*.

Before turning in Chapter 3 to just what these resources are and what a Humean version of scientific realism looks like, I pause in Chapter 2 to consider the textual evidence for and against this reading of Hume. I begin with a consideration of De Pierris's argument that Hume endorses what she calls a Newtonian inductivist understanding of scientific explanation, which amounts to something very much like DN. I find that the argument that she presents does not support this extreme conclusion, but only the more modest thesis that in order to form a theoretical representation we must employ a perceptible model. I then turn to Hume's own explicit statements concerning the nature of scientific explanation and show that it is only by approaching his texts with a bias towards understanding them as endorsing this radically impoverished view of explanation that one would read them as supporting that view. A more careful consideration of them reveals that what Hume objects to is not substantial explanation *per se*, but only such explanations that appeal to a *via negativa* with no descriptive content and therefore no explanatory power whatsoever. Following that, I review a plethora of examples in which Hume himself employs substantial explanation in just the way that I have argued he must. What these examples bring out is that Hume takes a theoretical representation to be the result of the combination of the deliverances of the senses and the recombinatory power of *reason*, which is a very surprising conclusion in need of its own investigation.

In fact, by the close of Chapter 2 there are several important pieces of my interpretation that need detailing, and so Chapter 3 pulls together and ties up these loose threads. Since I have argued up to this point that Hume endorses substantial explanation, explanation that appeals to the nature, powers, and essence of the substance underlying experience, my first order of business in Chapter 3 is to explicate Hume's understanding of substance, which I do via a close reading of 1.1.6 "Of modes and substance." He there argues that *contra* his predecessors the idea of substance is not the idea of some *via negativa* in which properties inhere and of which we can form no idea. Rather, the idea of substance is a general term with, "a uniting principle" as its "foundation." Working through the examples that he gives there of modes and substances, I conclude that a general term with a uniting principle as its foundation is a representation of certain manifest qualities as all stemming from a common cause. Because that cause is not itself directly represented, though, the puzzle from Chapter 1 concerning general representation again rears its head. How do we use a general term to represent the substance underlying manifest phenomena when the components of that general representation are all drawn from experience itself? In the second section of Chapter 3, I argue that since no resources *internal* to the mechanisms of general representation can do this work, it must be resources *external* to that theory that do so. Specifically, I argue that it is by *replacing* one set of general terms with another that we take the successor set to reflect a real ontological difference in the substance of the mind.

That mechanism accounts for our ontological commitment to the entities represented by the science of human nature, but not yet to the means by which such representations are made determinate. That is, replacing the language of the vulgar or the false philosophy with that of the scientist of human nature is the means by which we endorse the latter as an accurate representation of the substance of the human mind, but in order for that mechanism to operate, the scientist of human nature must first create a theoretical representation that determinately represents not just whatsoever explains the manifest phenomena, but also what it is in particular about that substance that does this explanatory work. To that end, the final section of Chapter 3 picks up on the discovery from Chapter 2 that *reason* plays some essential role in the formation of such theoretical representations.

In order to determine what precisely the role that reason plays in this regard is, I first locate it in Hume's taxonomy of cognitive faculties, which also includes memory and the imagination. Scholars have tended to focus their attention on the latter two faculties, debating whether these are distinguished by the degrees of force and vivacity of their product-representations or by the differing ways that each produces these representations (through reproduction and recombination, respectively). In such debates, little attention is paid to reason at all, and insofar as it is mentioned, it is usually as a sub-species of the imagination. I argue that the evidence should be clear that Hume takes the difference between memory and imagination to be that the former functions to reproduce the structure of impressions whereas the latter recombines ideas arbitrarily. I locate the source of the controversy in the secondary literature in a failure to pay close enough attention to the contexts in which Hume appears to draw this distinction via force and vivacity, which is unfailingly in distinguishing not these faculties themselves, but rather their influence on belief. As it turns out, a proper understanding of this distinction yields a place for reason as a distinct faculty after all. Specifically, I show that reason is, like the imagination, a recombinatory faculty, but like memory operates under the constraint provided by its distinctive function: reason's function is to *explain* the inductively-established universal regularities of experience. Thus, I argue that Hume countenances not only inductive (probable) reasoning and deductive (demonstrable) reasoning, but also a form of inference to the best explanation. Reason recombines the matter given in experience to form theoretical representations with novel content intended to explain the regularities that it discovers in experience.

With the basic elements of this Humean picture of scientific explanation in hand, in Chapters 4, 5, and 6, I develop it further by observing how Hume puts this understanding to work in his arguments concerning the ideas of external existence, necessary connection, and the self. What I argue in all three cases is that Hume's arguments are explicitly aimed at conclusions regarding the *representational content* of our ideas, specifically that the ideas that the vulgar and false philosophers take themselves to have in each of these cases is in fact impossible to form. This familiar take on Hume's arguments receives a new treatment in these three chapters. In its classical form what one would observe is Hume arguing that we can have no simple or complex ideas of such things. Given that I argue that Hume employs a third kind of representation, theoretical-explanatory posits, I take on the additional burden of showing that Hume also argues that we can form no theoretical representation of them either. Happily, that burden can be met because in all three cases there is ample evidence that theoretical representation is at the forefront of Hume's own thought.

In Chapter 4, I begin with Hume's arguments from 1.2.6 concerning the ideas of existence and external existence. I show that his conclusions are that we can form ideas of no such things, and that the common form that he attributes to both arguments consists of an appeal to the Representational Copy Principle, and to the conditions for theoretical representation. It is in presenting his conclusion that Hume alludes to what he calls a relative idea, which is the fulcrum of the New Hume interpretation, so in the second section of Chapter 4 I pause to demonstrate the untenability of Strawson's take on that passage in particular and on Hume's theory of representation and reference more generally. With that task completed, I turn to Hume's critique of the systems of philosophical explanation of the vulgar and false philosophy in 1.4.2 through 1.4.4. As Hume conceives them these competitor philosophical systems both attempt to form theoretical representations to explain the apparent constancy and coherence of our perceptions. That activity is not itself illegitimate *per se*, but each system nonetheless fails for exactly the reason that my interpretation of Hume's understanding of scientific explanation predicts. Specifically,

- Neither explanation is the result of *reasoning* about the phenomena at hand, but is rather a product of the imagination.

- Neither attempt at explanation is capable of specifying *determinate* differences and similarities between the model observable phenomena and the theoretical entity posited.
- The explanation of the vulgar straightforwardly contradicts empirical evidence.
- The explanation of the false philosophy, while it ostensibly replaces the language of the vulgar, is really entirely parasitic on it, i.e., only replaces the *words*, but not the ideas and/or associations that give these words their meaning.

Thus, Hume rejects both of these attempts to explain the apparent constancy and coherence of our perceptions. If, however, these explanations cannot explain those apparent inductively-established universal regularities of experience, then Hume appears to be left with an explanatory debt. In fact, though, what Hume himself discovers is that these are not universal regularities of experience at all. Rather, it is the illicit biases of the imagination as opposed to the reliable discoveries of reason that make it *appear* to us that experience is constant and coherent when in fact it is incredibly variable. Thus, Hume turns his attention to the only genuine *explicandum* in the vicinity—our regularly misconstruing this data—and explains this via an appeal to the nature, power, and substance of the human mind itself, its underlying composition and behaviors.

The subject of Chapter 5 is the idea of necessary connection, which provides an important proving ground for my interpretation in two ways. First, as with Hume's arguments concerning the idea of body, I begin by showing that Hume's conclusion is that we can have no idea of necessary connection as the vulgar or false philosophy conceive it, and that his arguments for this conclusion specifically address not only potential simple and complex ideas of necessary connection, but also theoretical representations of it. Second, though, it is not implausible to think that the notion of substantial explanation itself presupposes the idea of necessary connection, and so my rejection of that idea on Hume's behalf might seem to undercut the interpretation that I defend. Therefore, my second order of business in Chapter 5 is to give an account of substantial explanation that does not require necessary connection. That account begins with reconstruing the *aims* of substantial explanation. Rather than aiming to explain the regularities of experience via an appeal to the necessary connections underlying these, I argue that for Hume the proximate aim of substantial explanation is to render a given *explicandum* expected, or to form a "fixed and stable" expectation regarding its behavior. Forming such an expectation allows us to reliably track the behaviors of manifest objects, which is a means to achieving the distal aim of substantial explanation: to form an accurate representation that reflects the nature of the substance underlying the human mind.

Of course, it is not open to me to understand either of those goals as requiring an appeal to necessary connection, and so I next confront what appears to be a vicious regress. If an appeal to the nature of the substance underlying manifest phenomena cannot make use of the idea of necessity, then at most it will consist of an appeal to the regularities that this substance exhibits. Since, however, it is the regularities discovered in experience that demand explanation in the first place, the regularities attributed to substance will now appear to demand explanation in turn, etc. To address this concern I offer an interpretation of Hume's understanding of explanation as appealing to the *nature* of a thing according to which to attribute a property to the nature of a thing is to stipulate that that property will, in a given context, serve as an *explanans* and will not itself demand explanation *in that context*. Thus, while a potential regress looms, it is not a vicious one because the regress can be stopped in the context of any particular inquiry, and Hume considers all legitimate explanatory demands to exist only in the context of such particular inquiries. That a further explanation *can* always be demanded does not undermine the success of any particular explanation.

With those clarifications in hand, my account of Hume's understanding of scientific explanation is complete. I close Chapter 5 by returning to the puzzle posed in Chapter 1 to offer a detailed solution to it. I show that while we cannot form an idea that is an *image* of a theoretical posit such as simple ideas, what we can do is use a mixture of impressions and reason to form a set of general representations with determinate content that *reflects* what we take to be the true ontological structure of the substance of the human mind.

In the final chapter, I focus on Hume's recantation of his theory of personal identity in the Appendix, which has long puzzled scholars in large part because Hume is not explicit about what motivates that rejection. I argue that the most important clue that Hume does leave is that he twice voices dissatisfaction with his attempt to *explain the principles of connection* that govern the mind. The first important thing to notice about this way of phrasing his concern that it is the inductively-established universal regularities discovered in experience that are the *explicandum* of the account that Hume rejects. The second important thing is that what Hume offers in 1.4.6 is not an *explanation* of those regularities, but rather a mere description of them. In fact, 1.4.5 and 1.4.6 together constitute an argument that no such explanation is possible. That is, since Hume takes scientific explanation to consist in an appeal to the nature of the substance underlying manifest phenomena, in arguing that we can form no representation of the substance of the human mind—either as a soul or a subject-perception—what Hume argues is that no explanation of the unity of consciousness is possible. Since, however, Hume remains entirely satisfied with his description account of the mind, his delineation of the inductively-established universal regularities that it exhibits, he finds himself in a kind of methodological contradiction. On the one hand, there is a manifest phenomenon that demands explanation. On the other hand, he has demonstrated that no such explanation is so much as possible. Thus, he recants and retreats to the hope that someone else can discover a way forward.

That is the final piece of evidence in favor of my interpretation of Hume as a scientific realist. The first piece of evidence is the close examination of Hume's use of the simple-complex and impression-idea distinctions. It is to that study that I now turn.

Notes

1. Blackburn, "Hume and Thick Connections," frames the debate between Positivist Interpreters and New Humeans similarly.
2. For example, in Weinberg's nearly contemporaneous history of Logical Positivism, he cites Hume as being the Logical Positivist *par excellence*. Weinberg, *An Examination of Logical Positivism*, 3.
3. Sellars, "Language of Theories," 113–14.
4. Sellars, "Empiricism and the Philosophy of Mind," 167–8.
5. Kail, *Projection and Realism*, 78, justifiably rejects this moniker on the grounds that it carries with it the implication that such an understanding of Hume is neither Hume's original view nor one with a rich history in the secondary literature. Unfortunately, the use of 'New Hume' seems to have stuck, and so I will continue to use it, but it is worth an explicit disavowal of those implications. I hold that while the New Hume approach is wrong as an interpretation of Hume, as are most, it is nonetheless an entirely respectable contender worthy of serious attention. Kail also rightly points out that the views of the New Humeans differ in their motivations, arguments, and details. Some of these differences will be important farther along (mostly in Chapter 4), but can be glossed over for the moment.
6. Kail proposes an amendment to this thesis of Strawson's meant to forestall certain objections to it. He agrees with Strawson that we cannot have an idea with any descriptive content that represents powers, etc., but holds that we can represent these as, "that feature that, were we acquainted with it, would yield *a priori* inference and render it inconceivable that the cause not be followed by its effect" (Kail, *Projection and Realism*, 84). I will argue in Chapter 4 that this emendation is not enough to salvage the New Hume interpretation.
7. Kail, *Projection and Realism*, 56–7. I have rearranged Kail's text to match his definitions of each anti-realist position with their denominations. Also, while Kail's focus is on the external world and God, his definitions easily generalize.
8. So, for example, Kail's solution (which is similar to Garrett's) that certain beliefs can be justified by *practical* considerations if not theoretical ones is unavailable on my interpretive line.
9. Cf. Loeb, *Stability and Justification*, 22.
10. Garrett, *Cognition and Commitment*, 92.
11. Garrett, *Cognition and Commitment*, 234.
12. See for example Landy, *Kant's Inferentialism*.
13. Sellars, "Empiricism and the Philosophy of Mind," §41.
14. Sellars, *Science and Metaphysics*, §90.
15. Here is yet another set of issues in the secondary literature in which the current interpretation can be situated! In addition to doing so briefly in Chapter 1, and then revisiting the nature of specifically *theoretical* representations in Chapter 5, I also do so in more depth in Landy, "Recent Scholarship on Hume's Theory of Mental Representation."
16. (A revival set is roughly the set of resembling ideas that one recalls upon hearing a certain word.) 'Revival Set' follows Garrett, *Cognition and Commitment*.
17. This schema is one of those small debts to Sellars just mentioned (Sellars, "Naming and Saying," 207). He develops a point from Wittgenstein, *Tractatus Logico-Philosophicus*, §3.1432.

1 Two Case Studies

The Impression-Idea and Simple-Complex Distinctions

The strongest part of the case for understanding Hume's conception of the science of human nature as proceeding along the robustly realist lines that I will argue it does comes from a careful examination of his actual procedure in conducting that science. That is, while I will present evidence in the following chapters that Hume's explicit statements about science can be read as expressing his commitment to realism, that the case for reading Hume as an inductivist about scientific methodology is insufficient, and that all of his most important conclusions are reached via arguments that proceed using the methodology that I ascribe to him, the hard core of my interpretation consists of exposing the underlying logical structure of Hume's methodology in conducting his own investigation into the nature, powers, and essence of the human mind.

What I will argue in this chapter is that despite Hume's apparent commitment to a very strong form of nominalism about qualities and relations, his own pursuit of the science of human nature commits him to the reality of the qualities of and relations between the various kinds of perceptions that he takes to constitute the human mind. Specifically, I will show that both the impression-idea distinction and the simple-complex distinction are best understood as theoretical hypotheses about the real differences that underlie and explain the observed regularities of human thought.

My procedure here will be as follows. I will begin by briefly outlining Hume's theory of mental representation with an emphasis on the role of the Copy Principle in that theory and the surprising use to which it is put in general representation. Much of this material will likely be familiar, but its central importance to Hume's project, as well as a recent renewal of interest in the topic by Hume scholars, warrants review. I will then present a puzzle about the role of his theory of general representation in Hume's pursuit of the science of human nature, and specifically its compatibility with two of Hume's most important theoretical-explanatory moments: his drawing of the impression-idea distinction and the simple-complex distinction. Briefly, the puzzle is as follows. Hume appears to endorse a very strong version of nominalism—the thesis that only particulars, and not qualities or relations, exist. The distinctions that he employs in pursuing the science of human nature, however, only have the explanatory force that they do if they represent the real qualities and relations among our perceptions. Thus, I take Hume to be a kind of scientific realist: he is a nominalist about that which is represented by the general terms employed by the vulgar and false philosophy, but believes that those represented by the general terms of the science of human nature are in fact robustly real. The details of that solution to this puzzle and of this interpretation of Hume's understanding of the science of human nature will not be entirely worked out until the end of Chapter 5 because presenting it in its entirety requires understanding a great many of the intricacies of Hume's philosophical system, and how these all function together. Suffice it to say for now that what I will argue over the course of this study is that Hume employs a *three-tier* system consisting of a picture theory of mental representation, an account of reason's analogical extension of perceptible models into theoretical representations, and a linguistic mechanism of ontological commitment. We can begin with the first of these.[1]

Hume's Theory of Mental Representation

As is generally acknowledged, there is a great deal of textual support for the thesis that there is a very close link between Hume's notion of *copying* and his theory of mental representation.[2] The most straightforward way of drawing this link—defended by Cohon and Owen, "Hume on Representation"—is simply to identify them.

> Representation$_{CO}$: a perception represents only that of which it is a copy.

The key notion here is that of being a copy, and Hume is fairly clear about just what this entails. For x to be a copy of y requires that two conditions be met. These conditions are each necessary, and together are jointly sufficient, for x to be a copy of y. The first condition is that x must be caused by y. Of course, 'cause' must be construed in the proper Humean way here—more on this in Chapter 5—so that for x to be caused by y is for x and y to be constantly conjoined, and for y to always precede x. So, for example, when Hume sets out to prove the Copy Principle in the opening pages of the *Treatise*—the thesis that all simple ideas are copies of some simple impression—he observes that exactly these two parts of the causal condition are met.

> I first make myself certain, by a new review, of what I have already asserted, that every simple impression is attended with a correspondent idea, and every simple idea with a correspondent impression. From this constant conjunction of resembling perceptions I immediately conclude, that there is a great connexion betwixt our correspondent impressions and ideas, and that the existence of one has a considerable influence upon that of the other. Such a constant conjunction, in such an infinite

> number of instances, can never arise from chance; but clearly proves a dependence of the impressions on the ideas, or of the ideas on the impressions. That I may know on which side this dependence lies, I consider the order of their *first appearance*; and find by constant experience, that the simple impressions always take the precedence of their correspondent ideas, but never appear in the contrary order.
>
> T 1.1.1.8; SBN 4

Correspondent impressions and ideas are constantly conjoined, and the former always precede the latter. Thus, Hume can conclude that impressions are the cause of ideas (in the proper Humean sense of 'cause').

The second condition that must be met for *x* to be a copy of *y* is that *x* must exactly resemble *y*. Again, here is Hume in the opening pages of the *Treatise* offering evidence that this condition is met in the case of ideas and impressions.

> The first circumstance, that strikes my eye, is the great resemblance betwixt our impressions and ideas in every other particular, except their degree of force and vivacity.
>
> T 1.1.1.3; SBN 2

Of course, Hume goes on to limit his resemblance thesis to *simple* ideas and impressions only, and so correspondingly limits the Copy Principle to just these as well, which caveat will become important in a moment.

Before that, though, notice that what we have just been examining is Hume's defense of the Copy Principle, the thesis that all simple ideas are copies of some simple impression. That thesis is distinct from Representation$_{CO}$ insofar as the Copy Principle concerns only the causal relation between simple impressions and ideas and the relation of what we might call their *pictorial character*. It will be important to carefully distinguish these two principles. I will explain. Consider the following picture.

The *pictorial* character of this picture consists of four black lines of equal length arranged at ninety-degree angles to one another against a white background. That is in what *the picture* consists. For another picture to exactly resemble this one, it would also have to consist in four lines of this length arranged at ninety-degree angles to one another against a white background. The pictorial character of an image, including impressions and ideas, is constituted entirely by the *intrinsic* features of that image.

This is a point commonly made about Hume's exact resemblance thesis in order to explicate the notions of force and vivacity, which are not part of the pictorial character of perceptions precisely because they are not *intrinsic* features of a perception.[3] What concerns us here, however, is a slightly different contrast. Notice that in describing the image above, we made no reference to what that image *represents*. That is, we described the intrinsic features of that picture, but did not mention, for instance, that it is a picture

Figure 1.1

of a building as seen from directly above, or a book seen from straight on, etc. We can call what a picture is a picture *of* the *representational content* of that image, and it will be important for what follows to notice that it is, pre-theoretically, possible for the pictorial character of an image and the representational content of that same image to come apart. For instance, there are abstract paintings that certainly have *pictorial character* (a bunch of red, yellow, and blue paint splashes on a white canvas), but which do not have any *representational content* (these paintings are not paintings *of* anything). Conversely, in typical cases, the pictorial character of a written or spoken word is, in a sense, irrelevant to what it represents—hieroglyphics and onomatopoeia aside—(such words are not *iconic* representations; 'dog' does not look like a dog), but do have representational content (they are *representations*).[4] So, Hume's Copy Principle states that all ideas are copies of impressions, i.e., that all ideas are caused by, and have exactly the same *pictorial* character as, some corresponding impression. Important for us to note is that, as formulated here, the Copy Principle does not speak at all towards the *representational* content of impressions or ideas. It is merely a thesis concerning the causal relations between impressions and ideas and the relation of their pictorial character. $\text{Representation}_{CO}$ adds to the Copy Principle that not only are all ideas copies of impressions, but also that an idea *represents* only that of which it is a copy.[5]

As evidence for this understanding of Hume, Cohon and Owen cite Hume's use of the idiom of representation throughout his introduction and defense of the Copy Principle in the opening pages of the *Treatise* and the use to which he appears to put Representation$_{CO}$ in his arguments thereafter. In fact, though, as it stands, Representation$_{CO}$ cannot be right for reasons that Hume himself explicitly acknowledges, although a small change to it that preserves its spirit if not its letter will suffice to ameliorate the problem. As we just noted, Hume restricts the Copy Principle to simple impressions and ideas. The content of this claim from early on in the *Treatise* is only that every *simple* idea is copied from some *simple* impression.[6] Correspondingly, then, if it is only simple ideas that are copied from simple impressions, Representation$_{CO}$ must similarly apply only to simples. That is, since Representation$_{CO}$ states that a perception represents only that of which it is a copy, and only *simple* ideas are copies of anything, then it would seem to follow that only simple ideas represent anything. This, however, cannot be right.

First of all, any plausible theory of mental representation must to be able to account for the representational content of thoughts of complex items such as that of a dog, a person kicking a ball, etc. Hume clearly does so by casting such thoughts of complexes as themselves being complex ideas, and so he must have some account of how such ideas represent what they do. He certainly writes as if he does.

> I can imagine myself such a city as the *New Jerusalem*, whose pavement is gold and walls are rubies, tho' I never saw any such. I have seen *Paris*; but shall I affirm I can form any such an idea of that city, as will perfectly represent all its streets and houses in their real and just proportions?
>
> T 1.1.1.5; SBN 3

This is only one of a vast number of instances in which Hume indicates that he takes complex ideas to represent. This quotation also brings out a second problem with Representation$_{CO}$. If a perception represents only that of which it is a copy, then no perception can ever *mis*represent. If a perception represents only that of which it is a copy, then the object of any perception that represents at all is always guaranteed to exist (since it had to be copied in order to be represented), and so such representations can never go wrong. As we saw in the above quotation, Hume is clearly aware of the phenomenon of misrepresentation: we can represent New Jerusalem even though it has never existed, and our representation of Paris does not correspond exactly to the details of the actual city.

What is also indicated by the above quotation is that Hume takes these two problems to be connected: it is our ability to represent via complex ideas that leads to our making such errors. Recall the context in which Hume writes the above. He has just proposed for the first time that all of his ideas exactly resemble some impression. He then notices these two kinds of examples, and uses the distinction he has previously drawn between simple and complex ideas to restrict the scope of his claim to just simple ideas. The implication certainly seems to be that it is the fact that such ideas are complex that keeps them from exactly resembling some impression, and that this is not the case with simple ideas. Simple ideas cannot misrepresent; only complex ideas can.

Here, then, is a proposal for how to emend Representation$_{CO}$ to account for complex and misrepresentation, while keeping copying at the core of Hume's account. A simple perception represents only that of which it is a copy. Notice that a complex perception is nothing more than an arrangement of simple perceptions. So, take seriously the Humean slogan that a representation of a complex is nothing more than a complex of representations. That is, a complex perception represents the simple objects that are represented by its simple parts as being arranged in the way that those simple parts are arranged in it. Consider, for instance, Hume's account of the origin of our representations of spatial complexes.

> The table in front of me is alone sufficient by its view to give me the idea of extension. This idea, then, is borrow'd from, and represents some impression, which this moment appears to the senses. But my senses convey to me only the impressions of colour'd points, dispos'd in a certain manner. If the eye is sensible of any thing farther, I desire it may be pointed out to me. But if it be impossible to show any thing farther, we may conclude with certainty, that *the idea of extension is nothing but a copy of these colour'd points, and of the manner of their appearance.*
>
> T 1.2.3.4; SBN 34, emphasis mine

Our complex idea of a spatial complex comes to represent the spatial complex that it does by being a collection of simple ideas of colored points arranged in a way that exactly resembles the arrangements of the spatial complex being represented. We represent the relation that some simple impressions stand in to one another by arranging simple representations of each of these impressions into *the same relation*. We represent *a* as being next to *b* by placing an idea of *a* next to an idea of *b*. The idea of a spatial complex is nothing more than a spatial complex of ideas.

Hume is clear that our representation of temporal complexes works in the same manner.

> The idea of time being deriv'd from the succession of our perceptions of every kind, ideas as well as impressions, and impressions of reflection as well as of sensation, will afford us an instance of an abstract idea, which comprehends a still greater variety than that of space, and yet is represented in the fancy by some particular individual idea of a determinate quantity and quality.
>
> T 1.2.3.6; SBN 34

Our idea of time is "deriv'd from the succession of our perceptions." Hume's thought is that we represent two items as being related, now temporally, by placing them in a temporal relation to one another. That is, for example, we represent one thing as happening before another by having a representation of the former followed by a representation of the latter. So, whereas we represent a spatially complex state of affairs by forming a kind of picture before our mind's eye, we represent a temporally complex state of affairs by forming a kind of movie there. In general, complex ideas represent the simple impressions that their component simple ideas represent as being arranged in the way that those component ideas are arranged in the complex. We can generalize this account of complex representation using the general schema:

> 'x'R'y' represents xRy.

A representation of x and y as related to one another in way R consists of a representation of x related in the same way to a representation of y.

Notice that since these component simple ideas will be copies of the simple impression that they represent, when a complex idea *correctly* represents an arrangement of simple impressions, it too will be a copy of those impressions. That is, a complex idea is correct just in case it is caused by and exactly resembles the arrangement of simple impressions that it represents. An incorrect complex idea, on the other hand, such as Hume's ideas of Paris or New Jerusalem, will not be a copy of what it represents. Hume's idea of Paris does not exactly resemble it; his idea of New Jerusalem is not caused by any such city. Still, this necessary emendation to Representation$_{CO}$ maintains its spirit, if not its letter. Simple ideas represent that of which they are copies. Complex ideas represent those impressions that their simple components represent as being arranged as those ideas are arranged in the complex. They are correct when copies, incorrect when not.

Now consider Hume's theory of general representation and notice that general ideas are themselves complex ideas. What one might normally think of when one thinks of complex ideas are those that are a complex of simple ideas associated via relations of *contiguity*.[7] That includes our ideas of spatial and temporal complexes. A general idea, however, is a complex idea composed of ideas associated via relations of *resemblance*.[8] Thus, we should expect general ideas like more paradigmatic complex ideas to instantiate the schema just introduced of representing *x*R*y* via a representation of the form '*x*'R'*y*'. To see this we need first to have Hume's account in front of us, and it is brief enough to quote in full.

> When we have found a resemblance among several objects, that often occur to us, we apply the same name to all of them, whatever differences we may observe in the degrees of their quantity and quality, and

> whatever other differences may appear among them. After we have acquir'd a custom of this kind, the hearing of that name revives the idea of one of these objects, and makes the imagination conceive it with all its particular circumstances and proportions. But as the same word is suppos'd to have been frequently apply'd to other individuals, that are different in many respects from that idea, which is immediately present to the mind; the word not being able to revive the idea of all these individuals, only touches the soul, if I may be allow'd so to speak, and revives that custom, which we have acquir'd by surveying them. They are not really and in fact present to the mind, but only in power, nor do we draw them all out distinctly in the imagination, but keep ourselves in a readiness to survey any of them, as we may be prompted by a present design or necessity.
>
> T 1.1.7.7; SBN 20–1

When we encounter items that resemble one another in certain ways, we call them all by the same name. Because this name is frequently used in the presence of these items, our minds form an association between the two such that whenever that name is used we call to mind some one, or a few, ideas of these items. Furthermore, we stand disposed to call to mind more such ideas (ideas of the items between which this resemblance was found) upon further prompting. So, we represent persons in general by forming certain associations between our ideas of particular persons. Namely, those ideas all *resemble* one another, and are called to mind in the appropriate situations (primarily when the word 'person' is used, etc.) because of this resemblance.

One way to see that this account casts general representations as complex ideas is to ask what is represented by such general terms. The obvious answer is that, for instance, the general term 'personhood' represents personhood. That answer, however, simply pushes the question back a step: we now need to ask what kind of thing personhood is. The obvious first candidate here is persons.[9] That is, perhaps Hume's theory is that the general idea 'personhood' is just a way of representing all particular persons. Thus, the mechanisms that Hume details here are merely that: mechanisms for calling to mind just the ideas needed to represent all and only persons. On this reading of his theory, that we rely on the resemblance of each of these ideas to one another is an incidental feature of the account. If it just so happened that persons didn't resemble one another at all, but were encountered all standing next to one another, we would form the general idea personhood via associations of contiguous items rather than resembling ones.

This last point is implausible, though. First, such an idea of persons would be indistinguishable from an idea of all persons standing next to one another. Second, it does not seem to be a mere accident that it is resemblance that is at work in general ideas. In fact, a more plausible suggestion of what is represented by a general idea is exactly persons *as* resembling each other. That is, according to Hume's account, what prompts us to form general ideas in the first place is that, "we have found a resemblance among several objects." Thus, another natural suggestion for what is represented by an idea prompted by such encounters is that it is these objects *as* resembling one another. So, it is not just that our general idea 'personhood' represents persons, but more specifically, it represents persons as members of a set of objects among which we have found a resemblance. It represents persons *as* persons.

There is reason to think that Hume has just this kind of work in mind for general ideas. Here he is later in that section of the *Treatise* writing about distinctions of reason.

> Thus when a globe of white marble is presented, we receive only the impression of a white colour dispos'd in a certain form, nor are we able to separate and distinguish the colour from the form. But observing afterwards a globe of black marble and a cube of white, and comparing them with our former object, *we find two separate resemblances*, in what formerly seem'd, and really is, perfectly inseparable. After a little more practice of this kind, we begin to distinguish the figure from the colour by a distinction of reason; that is, we consider the figure and the colour together, since they are in effect the same and undistinguishable; but still view them in different aspects, *according to the resemblances, of which they are susceptible.*
>
> T 1.1.7.9; SBN 21–2, emphasis mine

Here Hume outlines how it is that we represent an object as being white as opposed to being a globe. What is represented is disambiguated by being represented *as* resembling some other object represented. This, however, is explicitly billed by Hume as being an instance of the use of general ideas. Thus, what is represented by a general idea is not just a group of particular objects, but rather a group of particular objects as resembling one another (and, thereby, as having a specific relation to one another).

Now, we have not yet seen Hume's account of complex representation play an explicit role here. It is, however, playing a role nonetheless. What we have tentatively established is that a general idea represents the items it represents as resembling one another. It is clear from Hume's account of general ideas that these function via the association of ideas of these items. That association is the one formed in virtue of these ideas resembling one another. So, what we have is that certain objects are represented as resembling one another in virtue of the fact that representations of these items are related to one another via their own relation of resembling each another. That is,

> 'x' resembling 'y' represents x resembling y.

General ideas are, in the end, an instance of complex representation. Objects represented are represented as resembling one another by placing representations of these objects into the same resemblance relations with each other. This is a surprising, but very tidy result. Hume's theory of how general ideas represent turns out to be an instantiation of his theory of how complex ideas represent. Here it is not spatial relations or temporal relations that are structuring the representations, but that ought to be expected since it is not spatial or temporal relations that are being represented. Rather, since it is resemblance relations that are being represented (persons are represented as resembling each other), it is resemblance relations that provide the structure. General ideas represent objects as resembling one another by placing representations of those objects into the same relation as the objects are represented as being in: resemblance relations. Hume holds that all representation proceeds via a suitably sophisticated version of the Representational Copy Principle.

'x'R'y' represents xRy.

Any representation that represents its object *as* having some feature does so by being a picture of that object, and what is pictured will depend on both the intrinsic features of the simple ideas at hand and the associative relations that structure these. This schema captures the most fundamental form that Hume's theory of mental representation takes. Hume holds that our perceptions represent in virtue of their being *pictures* or *images* of their objects. The elements of these pictures exactly resemble the elements of that which is pictured, and the relations that structure this picture represent those pictured elements as being arranged in the same relations. As we will eventually see, this picture theory of mental representation is at the very core of Hume's philosophical system. As we will see in the next section, though, it is also at the core of a significant puzzle about Hume's theory of general representation that threatens to undermine much of the explanatory force of and thus support for that philosophical system as well.

A Puzzle About Hume's Theory of General Representation

As we have just seen, according to Hume's theory of general representation, we represent generalities (such as dog-hood) by associating certain ideas with certain words (like 'dog'). On one prominent understanding of this theory, calling things by one name or another does not represent any real qualities of those things (their dog-hood) or any real relations between them (that they all resemble each other in some real way that they do not also resemble cats). Consider that theory again. Roughly, we find certain objects to resemble one another, and so apply a single word to all of them. Upon later hearing this word, we call to mind one or more of these objects. At the same time, we are disposed to recall other ideas that resemble these first ones when appropriately prompted. A particular idea thus made general represents the set of all of the ideas that we are disposed to recall upon hearing a certain word, and the meaning of the word, therefore, also extends to the entire set.

This account relies on the notion of our finding certain objects to resemble one another, and this "finding" can be taken in at least two ways. On the one hand, it might be that we discover the real relation that is capital-R Resemblance between these objects.[10] Call this the Ontological Interpretation. To understand Hume that way is to take him to be, in some sense, committed to an anti-nominalist thesis, namely, the thesis that there is at least one real relation: Resemblance. That is not a particularly attractive way to understand Hume given his announced nominalism, although some have argued that it is unavoidable.[11] Hume explicitly eschews the distinctive ontological commitment that the Ontological Interpretation ascribes to him. On the other hand, a line of interpretation is available according to which this "finding" of ideas to resemble one another is the not the discovery of a real relation between these items, but is instead nothing over and above our associating those items with each other and with the word (or general term) 'resemblance'.[12] Call this the Nominalist Interpretation. This association is no real relation, but differs only from other associations (e.g., of contiguous items, or of causes and their effects) insofar as we are inclined to associate our resembling ideas with one another, but not also with our contiguous ideas, etc. So, Hume's associations of ideas do not each represent a different real relation that ideas might stand in to one another, but rather only our inclination to parse our ideas into (roughly) three different groups. On this second and more *prima facie* attractive interpretive line, general terms turn out to be, in a sense, arbitrary. While they reflect the tendencies of the human mind to form associations between certain perceptions, this is all that they do. They do not, either in themselves or thereby, reflect any "deeper" reality.

As we will see in farther along, though, Hume *does* take at least some general representations—those that have a uniting principle as their foundation (T 1.1.6.3; SBN 17)—to represent the nature, powers, essence, or substance of the human mind. Furthermore, Hume's theory of general representation itself is essentially situated within his more comprehensive account of the workings of the human mind, crucial to which are certain key distinctions that Hume draws both right at the outset of, and throughout, the *Treatise*: the impression-idea distinction, the simple-complex distinction, the distinctions between memory, imagination, and reason, etc. According to the most thorough version of the Nominalist Interpretation of the theory of general representation, each of these distinctions is, again in some sense, arbitrary: they do not reflect any real relations or qualities of perceptions, but instead only how we happen to associate some, but not others, of these perceptions both with each other and with our names for them.

Here, then, is another puzzle. Each of these distinctions is meant to explain some phenomenon of the human mind. The simple-complex distinction, for example, is meant explain the possibility for novel human thought.

That explanation, however, has a distinctly, and arguably ineliminable, *ontological* component: complex perceptions are *composed of* simple ones. For the simple-complex distinction to do the explanatory work that it does in the *Treatise*, there must *really be* a distinction between simple and complex perceptions. The difference between these cannot be merely that some are associated with the term 'simple' and some with the term 'complex'. The latter must be composed of the former. The same will go for Hume's other core distinctions as well. Each is meant to explain some human phenomena; each plays a crucial role within Hume's philosophical system; and each has what will turn out to be an ineliminable ontological component that represents a real difference between the items associated with each term.

What I will argue here is that this is enough *prima facie* evidence for thinking that there is something wrong with the Nominalist Interpretation of Hume's theory of general representation *when it is applied to Hume's own theoretical-philosophical distinctions*. That is, I will argue that abandoning this strict form of nominalism is the form that Hume's scientific realism ultimately takes. While terms in the language of the vulgar must be understood purely nominalistically, the very explanation of why that (and much else) is the case demands that terms in the language of the scientist of man, such as 'impression' or 'idea', must be understood as robustly ontologically committing, as tracking real qualities of things and real relations between them. This realism, I will suggest, manifests itself in the form of Hume's replacing the language of the vulgar with the language of the scientist of man, and thereby endorsing the latter as a more accurate picture of the way that the world (or at least the world of perceptions) actually is.

While I will argue that carefully distinguishing these two modes of speech throughout the *Treatise* allows us to limit the range of the Ontological Interpretation to its appropriate sphere, even this requires a fairly radical departure from the Nominalist Interpretation. It requires us to understand Hume as working with a much more sophisticated conception of the science of human nature as has previously been attributed to him. That science is not a merely descriptive project, but a theoretical-explanatory one as well, which justifies the employment of ontologically committing theoretical posits.

My procedure will be as follows. The devil here is, as it usually is, in the details, specifically in the details of Hume's articulation and treatment of the particular distinctions at issue. So, my method for presenting this puzzle will be to focus on a pair of case studies: first of the simple-complex distinction, and then of the impression-idea distinction. My conclusion in each case will be that doing the explanatory work to which Hume puts them precludes understanding either of these distinctions as the Nominalist Interpretation would have it, but rather seems to require something more like the Ontological Interpretation. Working out the details of such an interpretation that is also consistent with Hume's professed nominalism and other philosophical commitments will constitute the matter of the next four chapters. Thus, the solution to this puzzle must be postponed until the close of Chapter 5. In the meantime, however, we can begin with a brief word about the kind of nominalism against which I will argue with respect to Hume's distinctions.

Humean Nominalism

As I understand the most strict version of Hume's nominalism, it is the thesis that the mind is constituted by nothing other than concrete mental particulars—perceptions—and that the so-called qualities and relations of these are all to be accounted for (or explained away) via appeals to the perceptions themselves and their behaviors. Specifically, according this strictest version of the Nominalist Interpretation, the account of general ideas in 1.1.7 is intended to provide a procedure for doing just this. A perception is blue not because it has the quality "blue," but rather because it is associated in the appropriate way with the word 'blue'. Two perceptions are related to one another, say as resembling, not because they stand in the "relation of resemblance," but rather because the complex idea of the pair of perceptions is associated in the appropriate way with the word 'resemblance'. This form of nominalism is admittedly extreme, but what I hope to accomplish in this section is to show that this is the form of nominalism to which Hume commits himself (at least with respect to the language of the vulgar), and that scholars attempting to interpret Hume as employing some less strict version of nominalism fail to do justice to Hume's texts and arguments.

The most canonical and concise statement of Hume's nominalism comes at T 1.1.7.6 where Hume writes, "every thing in nature is individual," but a great deal can be done using other parts of the philosophical system that Hume develops to understand what precisely Hume means by this claim, and to extrapolate his particular use of this thesis to other relevant issues. To begin, Hume famously uses this version of his thesis to argue, against Locke, that there can be no such thing as an "abstract" idea: an idea that is in some way itself indeterminate. While the Individuality Thesis, as we might call it, by itself, does not commit Hume to the very thoroughgoing nominalism attributed to him here—i.e., the thesis that the distinctions that we draw as described by Hume's account of general representation do not correspond to anything "real" in objects (or perceptions themselves)—what I will argue in this section is that it is plausible to take Hume to be committed to some version of that strong thesis nonetheless.

To begin, it is worth pausing to note that Berkeley, whose theory of general representation Hume cites as, "one of the greatest and most valuable discoveries that has been made of late years in the republic of letters," (T 1.1.7.1; SBN 17) is explicit in his application of this theory to relations, and Hume may well have simply carried over Berkeley's conclusion in this regard.[13] Hume's stated goal in that section of the *Treatise* is, "to confirm it [Berkeley's theory] by arguments, which I hope will put it beyond all doubt and controversy" (T 1.1.7.1; SBN 17). Hume here casts himself as adopting

Berkeley's theory and providing extra reasons in support of it. He does not announce a plan either to rehearse that theory in all of its detail or to amend it significantly. So, it would be unsurprising to find Hume discussing only the hard core of the theory, while omitting certain some of the less central elements e.g., its application to relations.[14]

Still, it is easy enough to construct an argument available to Hume from the Individuality Thesis to the conclusion that relations themselves cannot exist, using only premises that Hume more or less explicitly endorses. Here is one way:

(1) Everything in nature is individual (particular).
(2) Relations must be either particular or universal.
(3) If universal, relations do not exist. (1)
(4) If particular, relations must either exist independently of their relata, or must depend on their relata for their existence
(5) For the same reason that the abstract idea of a man cannot exist—either with the qualities of all men, or no qualities at all[15]—a relation cannot exist without relata. (Such a relation would have to either relate all of its relata, or none of its relata.)
(6) Thus, if relations exist, they are inseparable from their relata. (5)
(7) The Separability Principle: "whatever objects are different are distinguishable and [. . .] whatever objects are distinguishable are separable by the thought and imagination"

(T 1.1.7.3; SBN 18)

(8) Thus, if relations exist, they are not different from their relata. (6, 7)
(9) Just as the "whiteness" of a globe is really just the globe itself, relations between objects are just the objects themselves, i.e., only relata, and not relations, exist. (8)
(10) Relations do not exist. (3, 9)

Thus, combining Hume's stated nominalist thesis—the Individuality Thesis—with a version of the argument that he gives against "abstract" ideas, including his use of the Separability Principle, we arrive at precisely the view attributed to him above: that no relations exist. From this, we can derive the thesis of the Nominalist Interpretation as a specification of that principle: relations of resemblance between objects do not exist. So, in addition to the places in the Hume literature cited earlier where scholars have argued persuasively that Hume must hold some such view, we have here a straightforward argument from premises that Hume accepts that very view.

The motivation for taking this broad view of nominalism stems from the importance of the reductive nature of Hume's account of general representation. That is, if general representations consist of *only* what Hume describes, then supposing that our use of general terms corresponds to real facts is at best superfluous and at worst (if Hume follows Berkeley as I have suggested) contrary to the intended purpose of that account. Furthermore, consider that the theory of general representation that Hume presents in 1.1.7 accounts for our ability to form general representations without forming "abstract" ideas, ideas that are anything other than fully determinate particulars. The Nominalist Interpreter will point out that it would be odd for this elimination to be paired with, and in fact depend upon, an acceptance of the universals, relations, or real distinctions that would seem to be required by the Ontological Interpretation. That is, the Ontological Interpreter would have us eliminate abstract ideas, by relying on real relations of resemblance (or the corresponding real universals). While this is not inconsistent, it does at least seem to be an unhappy marriage.

Thus, I conclude that the Nominalist Interpretation is not only a live option for understanding Hume's position in the *Treatise*, but enjoys a number of important benefits over the Ontological Interpretation: it maintains consistency between Hume's declared nominalism and the conclusions that can be drawn using it, it is more in line with the spirit of Hume's typical order of explanation (wherein it is the science of *human nature* that is the *scientia mensura*), it can account for the relevant texts, and it avoids having to make a special exception for the treatment of the term "resemblance", which on the Ontological Interpretation requires, unlike any other term, an account other than the theory of general representation to explain it. From this I further conclude that the Nominalist Interpretation deserves to be privileged wherever it can be as how we understand the nature of the resemblance relation. If, however, the reader remains unconvinced of the viability of the Nominalist Interpretation, so be it. What I hope to show in what follows is that whatever one makes of this approach vis-à-vis the language of the vulgar, it will not suffice as an account of the general terms that Hume employs in his science of human nature. To that end, I will now turn to the first of two case studies: the simple-complex distinction.

The Simple-Complex Distinction

We will begin with Hume's distinction between simple and complex ideas. My argument here will be roughly as follows.

(1) One's phenomenology is always complex along multiple dimensions.
(2) Therefore, the revival set for the general term 'simple idea' will consist of complex ideas.
(3) Therefore, the general term "simple idea", must represent either (a) the complex ideas that constitute its revival set, or (b) something other than merely the members of its revival set.
(4) If the general term "simple idea" represents the complex ideas that constitute its revival set, then the simple-complex distinction loses its explanatory power (e.g., its role in explaining the novelty of human thought).

(5) If the general term "simple idea" represents something other than the member of its revival set, then the Ontological Interpretation of at least that distinction must be correct.

The more general conclusion that I will draw by the end of this section is that what Hume offers in the opening sections of the *Treatise* is a *theoretical scientific account* of experience. Our experience, the phenomena that a science of human nature must explain, is of only complex ideas. The nature of experience is explained in terms of posited simple perceptions that are governed by general laws and principles such as the Copy Principle, the laws of association, etc. This scientific theory of ideas should be contrasted with a descriptive phenomenology. If we restricted ourselves to experiential items that can be introspectively isolated, we would not appeal to simple ideas. Simple ideas are *posits*, or *theoretical entities*, because we have no direct experience of them as simple. It does, however, seem in line with an important strand in Hume's philosophy that simple ideas come to be known through a sophisticated scientific analogy with complex ideas. Hume's commitment to simple ideas is grounded not in direct experience of them, but in their *explanatory power*.

That is a fairly radical conclusion up to which we must work our way slowly. To begin our case study of the simple-complex distinction, then, consider the following. Human phenomenology, of the kind that Hume describes, is always and everywhere incredibly *complex*. Our visual field at any given time is a complex of spatial points; our diachronic experience is a complex of (themselves complex) temporal parts; our experience as a whole is a complex of perceptions delivered by our various sense modalities. The flipside of this observation about the ubiquitous complexity of human phenomenology is that that phenomenology is never *simple*. That is, it is never the case that one's phenomenology consists entirely of a single *simple* perception. One's visual field always consists of multiple perceptions arranged spatially; one's diachronic experience always consists of multiple perceptions *succeeding* one another; one's experience as a whole is always an experience involving perceptions delivered by more than one sense modality.[16]

Of course, it is one of Hume's central theses that this complexity is precisely a complexity composed of absolute simples: *minima sensibilia*. As we have seen, another important thesis of Hume's is that the meaning of any general term is the set of ideas associated with that term (in the way described in 1.1.7). And, of course, "simple idea" is itself a general term.

> And of this we may be certain, even from the very abstract terms *simple idea*. They comprehend all simple ideas under them. These resemble each other in their simplicity.
>
> T 1.1.7.7, fn. 5; SBN 635

Here, then, arises yet another puzzle: if our phenomenology is always complex, and the term "simple idea" derives its meaning from being associated with some set of ideas, then it would seem to follow that the ideas from which the meaning of "simple idea" derives are themselves *complex* ideas. That is, according to 1.1.7, the meaning of any general term is the set of ideas associated with that term. In the case of the general term "simple idea", since our phenomenology is always complex, this set must consist of *complex* ideas. And, in fact, this is exactly what we find in the *Treatise*. Consider, for example, one of Hume's proofs that some such *minima sensibilia* exist.

> Put a spot of ink upon paper, fix your eye upon that spot, and retire to such a distance, that at last you lose sight of it; 'tis plain, that the moment before it vanish'd the image or impression was perfectly indivisible.
>
> T 1.2.1.4; SBN 27

Consider carefully this procedure, and the image that it is meant to produce. One sees a spot on a piece of paper. That is a spatially complex perception. As one moves away from this piece of paper, the spot grows smaller and smaller until it disappears completely. That experience is both spatially *and* temporally complex. Presumably, one is hearing, smelling, seeing, etc., various "background noises" during this experience as well. What Hume's experiment presents, then, by way of demonstrating the existence of simple ideas is a spatially, temporally, and sense-modally *complex* set of experiences. So, insofar as this kind of procedure is what one is meant to recall upon hearing the term "simple idea", one is clearly meant to recall a complex *idea*.[17]

> Tho' a particular colour, taste, and smell are qualities all united together in this apple, 'tis easy to perceive they are not the same, but are at least distinguishable from each other.
>
> T 1.1.1.2; SBN 2

Here the simple ideas that we are meant to observe are the ones derived from the different sense modalities involved in a complex experience of an apple. All of those simple perceptions occur only as parts of the spatially, temporally, and sense-modally complex experience of biting into an apple. None is experienced *all on its own*, so to speak. So, in picking out such perceptions, "simple idea" cannot be straightforwardly associated with just simple ideas.

Of course, what one *wants* to say here is that while it may be certain complex ideas that we are meant to recall upon hearing the term "simple idea", it is the simple components of these ideas that we are picking out when we do so. We are meant to "focus on" the simple ideas to which these complex experiences "draw our attention". That may well be, but the question before us concerns what the *mechanism* is by which we are supposed to perform this operation. Hume offers a theory of general representation, and "focus" and "attention" are nowhere to be found in it. Insofar as a general representation as Hume describes it, "picks out" something over and above the set of ideas that one is disposed to recall upon hearing a certain term, such as the simple idea itself, then one has left the preferred Nominalist Interpretation well behind. It would, of course, be too hasty to draw this conclusion just yet.

Here one might also be tempted by the following line. While the revival set for "simple idea" will be composed of complex ideas, it is the simple components of these ideas (and not the complexes of which they are a part) that stand in the associative (resemblance) relations that are constitutive of this revival set. Thus, while the revival set is composed of complex ideas, it is not the composition of the revival set alone that fixes its content, but rather those of its components that stand in the proper associative (resemblance) relations to one another. Thus, because it is the simple components of the complex ideas that fix the content of the revival set for "simple idea", this revival set straightforwardly represents simple ideas as such.

Again, though, I believe that the framing problem concerning resemblance relations once again rears its head, and one must ask: in virtue of what do the resemblance relations among the complex ideas that one actually stands disposed to recall upon hearing "simple idea" hold between the simple components of these ideas rather than the complex ideas themselves? A revival set consists of a number of associated ideas, where "associated" means that upon recalling one of these ideas one stands disposed to recall the others. The particular association at work in revival sets is resemblance, and it is tempting to think, as above, that the ideas constituting the revival set all resemble each other in some way or other, e.g., in virtue of their simple components. What is at issue here, however, is exactly how to understand these resemblance relations. According to the distinctions that I have been employing, to suppose that the ideas constituting the revival set of "simple idea" resemble each other "in virtue of" their simple components, is to deploy the Ontological Interpretation rather than the Nominalist Interpretation because the former supposes that there is some relation of resemblance over and above the mere associations available via a descriptive phenomenology. Limiting ourselves just to such descriptions, we could only say that we recall, or stand disposed to recall, certain complex ideas upon encountering the words "simple idea". The temptation here is to suppose that there is something underlying such associations, something that explains that phenomenology. In fact, this is exactly what I will conclude is necessary by end of this chapter. Still, at this point in the dialectic, I am supposing that this is a conclusion that, given the arguments for the Nominalist Interpretation in the previous section, we should expect Hume to hope to avoid. So, our question is this: how can Hume rely on absolutely simple ideas to explain experience when we experience only complex ideas, and even the term "simple idea" itself appears to pick out a revival set of only *complex* ideas?

The key to answering this question is to see that simple ideas are theoretical posits. Hume's "science of man" (T Intro. 4; SBN xv) is meant to explain various phenomena. And like the natural sciences, Hume's science of human nature will employ in its explanations theoretical objects that are themselves not phenomena and not "directly" experienced. Hume posits such entities on the grounds that they *explain* certain phenomena. In particular, Hume posits simple ideas on the grounds that they explain how it is that experience alone can account for the complex ideas that we in fact have. This strategy can be seen as being introduced in the opening passages of Treatise that we have already begun to discuss. In these opening paragraphs, Hume is aiming at the first, crucial, theoretical proposition, the Copy Principle: "that all our simple ideas in their first appearance are deriv'd from simple impressions, which are correspondent to them, and which they exactly represent" (T 1.1.1.7; SBN 4). He first motivates the Copy Principle by affecting to notice that all of his ideas exactly resemble some impression he has experienced.

> The first circumstance, that strikes my eye, is the great resemblance betwixt our impressions and ideas in every other particular, except their degree of force and vivacity.
>
> T 1.1.1.3; SBN 2–3

As we have seen, upon further reflection Hume notices that he has many ideas of which this appears to be untrue, offering as examples the imagined New Jerusalem and inaccurately-remembered Paris.

> Upon a more accurate survey [. . .] I observe, that many of our complex ideas never had impressions, that correspond to them, and that many of our complex impressions are never exactly copy'd in ideas.
>
> T 1.1.1.4; SBN 3

The resemblance between ideas and impressions struck him as a plausible, naive beginning to an account of the origin of these ideas. In fact, once these apparent counterexamples are dealt with, Hume will go on to use the exact resemblance of simple ideas to simple impressions to do an enormous amount of work in the remainder of the *Treatise*. So, already treating the Copy Principle as true, he suggests that our mental architecture must be more nuanced than one might suppose pre-theoretically. In particular, he now puts to work the distinction between simple and complex ideas that he introduced a few paragraphs previously.

> After the most accurate examination, of which I am capable, I venture to affirm, that the rule here holds without an exception, and that every

simple idea has a simple impression, which resembles it; and every simple impression a correspondent idea.

T 1.1.1.5; SBN 3

In line with the interpretation of Hume as realizing that simple ideas are not experienced "as such", this passage should not be read as saying that he confirms the Copy Principle by an "accurate examination" of simple ideas "in isolation" along with copied impressions that are similarly experienced on their own. Hume is instead noticing that it will be virtually impossible to find a counterexample to the thesis that the imagination can construct complex ideas from postulated simple ideas that are themselves always copied from correspondingly postulated simple impressions.

Of course, because simple ideas *constitute* complex ideas, there is a sense in which all experience is "of" simple ideas. This sense in which we do experience simples must, however, be brought into sharper contrast with the sense in which we never experience simple ideas "as such." Consider, for example, that a pure sample of mercury in a beaker consists of nothing but mercury atoms. Now, when we look at this mercury or plunge a finger into it and feel it, we certainly do not have visual or tactile experience of individual mercury atoms "as such". This is so despite the fact that, by assumption, we are seeing or feeling nothing but the mercury atoms insofar as we see or feel the sample of mercury. We experience the mercury as a silvery cold liquid, but (especially before learning about atomistic theory) not as a complex substance consisting of billions of individual atoms of atomic weight 80. The vial mercury *consists of* mercury atoms, but in seeing or touching it, we do not experience it *as* a collection of mercury atoms. Analogously, while Hume's science of human nature teaches us that our phenomenology *consists of* a large number of simple ideas, we do not experience this phenomenology *as* a collection of simple ideas. Also analogously, just as we do not typically, or ever at all, experience single individual mercury atoms "in isolation," but come to posit these as explanations of certain observed phenomena, we also do not ever experience simple ideas "in isolation," but come to posit these as explanations of certain observed phenomena (in this case, certain psychological phenomena).[18]

One might be tempted here to draw an important *dis*analogy between the vial of mercury and Hume's phenomenological explorations on the grounds that while we might be wrong about the composition of the mercury, we cannot similarly be wrong about the composition of complex ideas because the latter are available to us through introspection and in this way present themselves *as* the simple elements of our complex ideas. The most outstanding text in this regard is the following passage from 1.4.2.[19]

> For since all actions and sensations of the mind are known to us by consciousness, they must necessarily appear in every particular what they are, and be what they appear. Every thing that enters the mind, being in

> *reality* a perception, 'tis impossible any thing shou'd to *feeling* appear different. This were to suppose, that even where we are most intimately conscious, we might be mistaken.
>
> T 1.4.2.7; SBN 190

This passage certainly does seem to support the suggestion that introspection affords us infallible access to the nature of our own perceptions. "[T]hey must necessarily appear in every particular what they are, and be what they appear." There is no room for us to be in error about, or to be unaware of, any fact about our perceptions.[20] Thus, it appears that my thesis that simple ideas are theoretical posits cannot be right because it would seem that introspection itself gives us direct access to our perceptions *as simple*. If not, "where we are, most intimately conscious, we might be mistaken."

If what I have been arguing thus far is correct, this line of thinking has to be a mistake, and I believe that there is much in the *Treatise* to recommend against it, but one example will have to suffice to make the point in the current context. Suppose that we have infallible access via introspection to the precise and complete nature of our simple ideas as simple. What, then, are we to make of Hume's *argument* that our perceptions are not infinitely divisible?

> 'Tis universally allow'd, that the capacity of the mind is limited, and can never attain a full and adequate conception of infinity: And tho' it were not allow'd, 'twou'd be sufficiently evident from the plainest observation and experience. 'Tis also obvious, that whatever is capable of being divided *in infinitum*, must consist of an infinite number of parts, and that 'tis impossible to set any bounds to the number of parts, without setting bounds at the same time to the division. It requires scarce any induction to conclude from hence, that the *idea*, which we form of any finite quality, is not infinitely divisible, but that by proper distinctions and separations we may run up this idea to inferior ones, which will be perfectly simple and indivisible.
>
> T 1.2.1.2; SBN 26–7

I believe that it is uncontroversial that, for Hume, to conceive of a perception as simple is to conceive of it as an indivisible *minimum*. If, however, we have infallible access via introspection to the precise and complete nature of our simple ideas as simple, then no argument should be necessary to demonstrate that our perceptions are indivisible. We ought to be able (infallibly) to introspectively perceive that they are simple and indivisible. As above, though, Hume does offer an argument for that conclusion, and that argument does not make use of our introspective evidence at all.

In fact, Hume uses this argument, rather than any introspective evidence to show not only that the products of imagination are only finitely divisible, but also that those of the *senses* are as well.

> 'Tis therefore certain, that the imagination reaches a *minimum*, and may raise up to itself an idea, of which it cannot conceive any subdivision, and which cannot be diminished without a total annihilation. [. . .] 'Tis the same case with the impressions of the senses as with the ideas of the imagination.
>
> T 1.2.1.3–4; SBN 27

As the "therefore" indicates, Hume takes the argument from 1.2.1.2 to be the evidence for his claim about the nature of the imagination, and he explicitly links this same evidence to his claim that impressions are likewise indivisible. So, neither the imagination nor the senses appear to represent simple ideas as simple, much less infallibly so, thus making necessary Hume's appeal to the original argument.[21] Combine this example with all the others throughout the *Treatise* of Hume's pointing out the *errors* that his predecessors have made in their theorizing about human nature, and one must conclude that Hume does not, in fact, take introspection to be an infallible, precise, and complete source of information about the human mind. Such examples are only possible on the presupposition that the deliverances of introspection, which his predecessors surely also employed, is fallible. We can be, and often are, mistaken about what it is that goes on in our minds. The hypothesis that our phenomenology is a complex composed of simples is a substantial scientific/philosophical thesis, in need of evidence and defense. This would not be the case were it possible to introspect and thereby become infallibly aware of the truth of this supposition. Thus, one should not suppose, merely on the grounds that Hume is seeking to explain our phenomenology, that everything in that explanation will either likewise be accessible to introspection, nor that even if it is, we can know it infallibly.

What then of Hume's declaration in 1.4.2.7 that all sensations, "must necessarily appear in every particular what they are, and be what they appear"? What I would suggest is that this is one of many places where Hume quickly tries a line of argument to which he is not really wedded, and that we ought not to hold him to this particular instance given how much tension there is between that thesis and so much else in the *Treatise*. As evidence for this reading of that passage, notice that in the paragraph immediately following it, Hume abandons the reasoning there as being open to too many objections, and turns instead to a more empirically informed (and thus more in line with his professed methodology) inquiry.

> But not to lose time in examining, whether 'tis possible for our senses to deceive us, and represent our perceptions as distinct from ourselves, that is as *external* to and *independent* of us; let us consider whether they

> really do so, and whether this error proceeds from an immediate sensation, or from some other causes.
>
> T 1.4.2.8; SBN 190

Hume recognizes the difficulty in maintaining the thesis that it is not, "possible for our senses to deceive us," and so abandons his argument based on that premise to pursue the question, not of the *possibility* of deception in the case of the idea of distinct existence, but rather that of whether the senses are *actually* the cause of error there. Clearly, while Hume thought it worth mentioning the more *a priori* reasoning about the infallibility of introspection, it is not a line to which he strongly committed, and thus given its conflict with so much else in his system, we ought not hold him to it. Thus, the supposed infallibility of introspection is not an obstacle to understanding simple ideas as theoretical posits because Hume is not committed to any such thesis.

Of course, positing simple ideas does no *explanatory* work until one says something about their nature. In fact, Hume introduces simple ideas in much the same way that other scientists introduce their own theoretical posits, and as we have seen his explicit statements about the science of human nature require. He introduces a perceptible *model* for simple ideas, and then describes how the posit both resembles and differs from the model. In Hume's case, the model is *a complex idea* that is experienced in the normal way. Here again is the passage in which Hume introduces the distinction.

> Simple perceptions or impressions and ideas are such as admit no distinction nor separation. The complex are contrary to these, and may be distinguish'd into parts. Tho' a particular colour, taste, and smell are qualities all united together in this apple, 'tis easy to perceive they are not the same, but are at least distinguishable from each other.
>
> T 1.1.1.2; SBN 2

Simple ideas are much like complex ideas; most importantly, they are mental images. The main difference is that the complexes but not the simples can be distinguished into parts. As we see later on in the *Treatise*, this model is enough to do heavy lifting for Hume. It specifies how simple impressions are copied by simple ideas. It also explains how complex ideas can come to have the structure they do. It will also play a role in accounting for the content of complex ideas: they are composed of simple ideas; their content derives from the simple ideas. And, as already noted, we have an explanation of how complex ideas often, but not always, resemble complex impressions: simple ideas resemble simple impressions, but can be recombined to form novel complex ideas.

All of this amounts to a *theoretical scientific account* of experience. Our experience, the phenomena that a science of human nature must explain, is of only complex ideas. The nature of experience is explained in terms of posited simple perceptions that are governed by general laws and principles such as the Copy Principle, the laws of association, etc. This scientific theory of ideas should be contrasted with a descriptive phenomenology. If we restricted ourselves to experiential items that can be introspectively isolated, we would not appeal to simple ideas. To repeat, simple ideas are *posits*, or *theoretical entities*, because we have no direct experience of them as simple. It does, however, seem in line with an important strand in Hume's philosophy that simple ideas come to be known through a sophisticated scientific analogy with complex ideas. Hume's commitment to simple ideas is grounded not in direct experience of them, but in their *explanatory power*.

We are now in a position to return to our puzzle about general representation. On the one hand, we can understand the simple-complex distinction as the Nominalist Interpretation has it: as merely reflecting our custom or habit of associating some perceptions with the term "simple" and other with the term "complex". On the other hand, we can understand it as the Ontological Interpretation has it: as reflecting a real distinction between two really different kinds of perceptions, in this case, those that can be distinguished into parts, and those that cannot. As predicted, the preferred Nominalist Interpretation appears to run into some fairly serious trouble here. In particular, Hume posits simple ideas for the sole purpose of gaining certain *explanatory power*, but that power is lost if one supposes that simple ideas are *just* those "complex" ideas that are associated with the term "simple idea".

Notice first that the key thesis here—that complex ideas are composed of simple ideas—on its most natural reading, comes out obviously false on the Nominalist Interpretation. Complex ideas are not composed from, in any recognizable sense of that word, the particular complex ideas that are associated with the term "simple idea". Recall that the revival set of the term "simple idea" will include in it various complex ideas, the paradigms of which are the dot on the piece of paper and the complex sense experience of biting into an apple. If we take the Nominalist Interpretation seriously, Hume's thesis would have to be that all of our ideas are composed of ideas resembling these. In fact, though, the Ontological Interpretation has an even stronger case to make here. As we noted earlier, the revival set for "simple idea" is composed entirely of certain complex ideas. The Ontological Interpreter will argue that what is needed, given the fact of the complexity of human phenomenology, is precisely a way to move *beyond* what can be captured in a revival set. What is needed, the Ontological Interpreter will insist, is a way of understanding Hume's theory of general representation according to which the associations that are involved in using a general term can (or can fail to) capture some *real relations* of the objects represented. On this line, while it is true that what we *associate* with the general term "simple" is a certain set of complex ideas, what is captured by, or reflected in, or represented by, this association is the *real difference* between simple and complex ideas.[22]

Let me spell out this line a bit more. Beginning with Hume's assertion that the difference between simple and complex perceptions is that the latter, but not the former can be distinguished into parts, we can take the revival set of "simple idea" to be a set of ideas in which this distinguishing manifestly fails. So, for instance, consider again the example of the disappearing dot. One can, again using the theory of general representation, distinguish the parts of the scene imagined, and in particular the dot from its background by forming a revival set consisting of, for example, ideas of similar dots against different colored backgrounds, etc. Because the dot itself, however, is a *minima sensibilia*, one cannot similarly distinguish any of its parts from one another. The revival set for "simple idea" will consist of many instances of such distinguishing and failures to distinguish. The question before us, then, is how to understand what is represented by such revival sets. If the Nominalist Interpretation is right, the answer is nothing. These acts of distinguishing and failing to distinguish represent nothing real, but must be understood as merely a (rather perverse) subset of the associative tendencies and customs of the human mind. Philosophy makes us do weird things, and this is one them.

Something about that last sentiment certainly rings true for Hume, but as I've been trying to emphasize, endorsing it comes at a steep price. Consider finally the explanation itself that the Nominalist Interpreter must see Hume as offering. The question before Hume is how to explain the great novelty of human thought, given the *prima facie* plausibility of the thesis that all ideas are derived from impressions. According the Nominalist Interpretation the answer to this question is that all perceptions associated with the term "complex" are "composed of" perceptions associated with the term "simple". As we noted earlier, this is simply and obviously false. What is worse, though, is that it robs Hume's response of all of its explanatory power. The power of the simple-complex distinction lies in its ability to make easy sense of compositionality, decompositionality, combination, and recombination. The Nominalist Interpretation, by understanding "simple" as, as we might put it, complex-but-associated-with-"simple", undermines all of those resources. It is not enough that we be come to associate some perceptions with "simple" and others with "complex". For this explanation to work, those associations *must* reflect some real properties of the perceptions so associated: namely their simplicity, complexity, and the fact that the latter are composed of the former. The mere associations of certain but not other of our perceptions with the terms "simple" and "complex", while it might explain something, in no way explains that which the simple-complex distinction is meant to explain, namely, the novelty of human thought. Again, the mere fact that we associate some perceptions with the word "simple" and others with the word "complex" does not explain how it is that we can create novel ideas from the limited store provided by experience.

So it seems that understanding Hume's simple-complex distinction using the Nominalist Interpretation of the theory of general ideas undermines the explanatory role that is the primary purpose of introducing that distinction.

As we are about to see, this is not a feature particular to the simple-complex distinction. It causes a similar problem with the impression-idea distinction, and generalizes to all of Hume's most important theoretical apparatuses.

The Impression-Idea Distinction

In order to begin our study of the impression-idea distinction we must first settle on a single understanding of what that distinction actually is (although, as we will see, the puzzle we are considering would arise equally for alternative interpretations of the distinction as well). For many years, there was fairly widespread consensus in the secondary literature, insofar as the issue was considered at all, that the constitutive difference between impressions and ideas is, for Hume, that the former are more forceful and vivacious than the latter. Hume does, after all, open the *Treatise* by drawing this distinction in what appears to be just that way.

> All the perceptions of the human mind resolve themselves into two distinct kinds, which I shall call IMPRESSIONS and IDEAS. The difference betwixt these consists in the degree of force and liveliness, with which they strike upon the mind, and make their way into our thought or consciousness.
>
> T 1.1.1.1; SBN 2

Prompted by Barry Stroud's famous critique of this way of understanding the distinction,[23] scholars have recently explored alternative renderings of it with varying degrees of success. My own proposal has been that the difference consists in the fact that whereas ideas are *copies* of some other mental entities (impressions), impressions are, by contrast, mental *originals*.[24] While we talk about impressions and ideas as feeling and thinking, and while we recognize each using their degree of force and vivacity, the real difference between them, what *grounds* this talk and this recognition, is that whereas ideas are copies of some other mental entity, impressions are mental originals. My plan is to proceed by accepting this line on the impression-idea distinction, and seeing what insights it yields with respect to the place of this distinction in the broader context of the theory of human nature that Hume presents in the *Treatise*. To do that, it will be helpful to walk through the opening pages of the *Treatise*, where this distinction is first introduced, and to see how this line portrays Hume's development of it.

As we should expect, having peeked ahead to the theory of general representation, Hume begins his articulation of the impression-idea distinction by priming his reader to form the appropriate revival sets for the terms "impression" and "idea". To do that, he needs to draw his reader's attention to those salient qualities of each kind of perception that will cause the reader to form the associations of resemblance linking impressions on the one hand, and ideas on the other. We have already seen Hume taking the first step in doing this, drawing his reader's attention to the phenomenal qualities by which each can be recognized: their degree of force and vivacity. Next, Hume offers up paradigm examples of each, which are meant to bring into focus the resemblance relations that he hopes his reader will form.

> under this name ['impressions'] I comprehend all our sensations, passions and emotions, as they make their first appearance in the soul. By *ideas* I mean the faint images of these in the thinking and reasoning; such as, for instance, are all the perceptions excited by the present discourse, excepting only those which arise from the sight and touch, and excepting the immediate pleasure or uneasiness it may occasion.
>
> T 1.1.1.1; SBN 2

Next, Hume relates the new distinction that he is drawing to what he takes to be one that is antecedently and pre-theoretically familiar to his reader.

> it will not be very necessary to employ many words in explaining this distinction. Every one of himself will readily perceive the difference betwixt feeling and thinking.
>
> T 1.1.1.1; SBN 2

Finally, Hume contrasts his own idiom to that of one of his important predecessors.

> Perhaps I rather restore the word, *idea*, to its original sense, from which Mr. *Locke* had perverted it, in making it stand for all our perceptions.
>
> T n2; SBN 2

Despite what his opening force-and-vivacity idiom suggests, then, Hume has *not yet* actually drawn the distinction between impression and ideas, and has not yet told us in what, if anything, this distinction consists. Thus far, the only important philosophical work that Hume has done is to have primed his reader to begin forming the associations of resemblance between what he will call "impressions" and "ideas".

A proponent of the Nominalist Interpretation might be tempted to think that the story ends there. Having gotten his reader to form these associations, there is nothing more that Hume can do to draw this distinction because the theory of general representation is explicit that *this is all there is* to such terms. It is odd, then, to find Hume, after also introducing the simple-complex distinction, writing the following.

> Having by these divisions given an order and arrangement to our objects, we may now apply ourselves to consider with the more accuracy their qualities and relations.
>
> T 1.1.1.3; SBN 2

Qualities? Relations? If *all* that it is to be an impression is to be associated with the term "impression", etc., and *all* that it is to be an idea is to be associated with the term "idea",' etc., then there can be no good reason for thinking that either all impressions or all ideas will have any other non-trivial qualities or relations in common.[25] That is, if the theory of general representation is to be understood as the Nominalist Interpretation would have it, the categorizing of our perceptions into impressions and ideas is arbitrary and without any corresponding ontological underpinning. If that is the case, however, then Hume should have absolutely no reason to expect that impressions and ideas will have *any* qualities or relations in common other than the "quality" or "relation" of being associated with the terms "impression" and "idea", etc. The very essence of the Nominalist Interpretation is that our "finding" certain objects to resemble one another does not reflect anything robust about those objects: it is merely a quirk of the associative tendencies of the human mind.

And yet, Hume's behavior here indicates that this is precisely not how he understands at least this distinction. He clearly expects to discover—and takes himself successfully to discover—that impressions share certain interesting qualities, that ideas share certain interesting qualities, and that they each bear certain interesting relations *as a kind* to other kinds of perceptions. Our question, then, must be why Hume might find himself with this expectation, and what precisely he takes himself to subsequently discover. To answer the first question, it will be helpful to complete our survey of the opening of the *Treatise* because these pages clearly contain at least a superficial answer to the second question.

So, back to Hume.

> The first circumstance that strikes my eye, is the great resemblance betwixt our impressions and ideas in every other particular, except their degree of force and vivacity.
>
> T 1.1.1.3; SBN 2

After making the appropriate modification to this observation using the simple-complex distinction, this becomes the Exact Resemblance Thesis: that every simple idea exactly resembles some simple impression. That is significant because the exact resemblance is the first of two necessary and jointly sufficient conditions for something's being a *copy* of something else. The second, of course, is that the copy must be *caused by* that which it exactly resembles. As is familiar, this causal relation itself is divisible into two further conditions: the cause must precede the effect, and the cause and effect must be constantly conjoined.

> every simple impression is attended with a correspondent idea, and every simple idea with a correspondent impression. [. . .] I consider the order of their *first appearance*; and find by constant experience, that the

> simple impression always take the precedence of their correspondent ideas, but never appear in the contrary order.
>
> T 1.1.1.8; SBN 4–5

Thus, Hume is able to establish his famous Copy Principle, the thesis that,

> All our simple ideas in their first appearance are deriv'd from simple impressions, which are correspondent to them, and which they exactly resemble.
>
> T 1.1.1.7; SBN 4

So, perhaps the single most important relation that Hume discovers that impressions and ideas enter into is that the latter are all copies of the former. As I suggested at the outset of this section, as I understand Hume, this straightforward matter of fact about impressions and ideas will go on to do even more work for him than is normally supposed. In particular, what I have argued elsewhere is that upon discovering the Copy Principle, Hume subsequently takes this matter of fact to be what *grounds* and *explains* the distinction between impressions and ideas. That is, what Hume discovers in "considering with the more accuracy the qualities and relations of impressions and ideas" is that what it is to *be* an idea is to be a copy of some other perception, and that what it is to *be* an impression is to be a mental original.

To summarize, then, Hume begins by announcing that he wants to call his reader's attention to a certain distinction, one with which they are likely antecedently familiar (in the form of thinking and feeling), and which it should be easy for them to draw (using degree of force and vivacity as its indicator). On the interpretive line that we have been using, he then presents what he takes to be the real underlying difference between these two kinds of objects. While we talk about impression and ideas as feeling and thinking, and while we recognize each using their degree of force and vivacity, the real difference between them, what *grounds* this talk and this recognition, is that whereas ideas are copies of some other mental entity, impressions are mental originals.

An analogy to a familiar scientific example can help bring out the sense in which Hume seems to be committed to a robust ontological conclusion here, despite being concerned primarily with what appears to be merely phenomenological. To understand the scientific-explanatory dialectic in which the impression-idea distinction is situated, consider the discovery that gold and lead have different atomic weights. Pre-scientifically, we were very good at telling the difference between gold and lead. We could point to certain qualities that made these two kinds of substance easy to tell apart (e.g., their colors). We could present examples of each and expect others to extrapolate from these accordingly. When we discover that gold and lead have atomic weights of 79 and 82, respectively, however, we discover what *grounds* the pre-scientific distinction that we drew between them. What the discovery of atomic weights did was to ground our pre-scientific distinction in some real ontological one. Gold and lead are *composed of* sub-atomic particles, these particles have such-and-such properties, and that is what makes gold and lead really different. It is the exact same explanatory process that is at work in Hume's impression-idea distinction: impressions and ideas *are* mental originals and copies, respectively, originals and copies have such-and-such properties, and these are what *make* impressions and ideas different.

That is what Hume takes himself to discover. The question we must answer now is how we are to understand this discovery. By what right can Hume take himself to have hit upon certain qualities and relations that provide grounds for anything else?

If we consider just the first few pages of the *Treatise*, then the justification for taking ideas to be copies and for impressions to be mental originals is simply that doing so successfully explains our (mostly correct, but occasionally incorrect, as in "madness or fever") pre-theoretical use of terms such as "thinking" and "feeling", and our inclination to sort perceptions according to their degree of force and vivacity. For instance, it will turn out to be because ideas are copies and impressions originals that the former are less forceful and vivacious than the latter, and it is that fact, in turn, that helps us introspectively distinguish one kind of perception from the other as we do. (The gist of this explanation is that when ideas are copied from impressions some, but not all, of the force and vivacity from the latter is transferred to the former.)

Already here we can see the distinction between impressions and ideas—qua the distinction between mental originals and copies—playing a crucial explanatory role, which it would be difficult to separate out from its attendant ontological commitment. That is, if the *only* difference between being an impression and being an idea were the term with which a perception was associated, then explaining our association of those perceptions with those terms, as Hume seems to do, would be impossible. Imagine how that explanation would have to go. We find ourselves associating certain perceptions with the term "impression" (or "feeling") and others with the term "idea" (or "thinking"). We wonder whether there is anything that underlies our doing so. We discover that what we have been calling "ideas" are all copies of what we have been calling "impressions", etc. That seems explanatory until we additionally discover that what it is to be a copy is merely to be associated with the word "copy", which in turn is to be associated with the words "resemblance" and "cause". And so on for any term that comes to play a role in any such explanations. All are cast by the Nominalist Interpretation as *merely* reflecting the associative tendencies of the human mind. If, however, what we are seeking is an explanation of these very tendencies, then surely any such explanation is circular and therefore a failure.

This brings us to the real reason that Hume takes himself to have hit upon certain qualities and relations that provide the grounds of the impression-idea distinction: that distinction affords enormous explanatory power, which power is only gained by accepting a real distinction between impressions and ideas *contra* the Nominalist Interpretation of the theory of general representation. Consider the implications of accepting the Nominalist Interpretation of this distinction. To be an impression would be *just* to be associated with the word "impression", etc. To be an idea would *just* to be associated with the word "idea". To claim that impressions are more forceful and vivacious than ideas would *not* be to claim that there is an introspectable quality that impressions have to a higher degree than ideas. Except, that is, insofar as what we meant by this claim is that that those perceptions associated with "impression" are also associated with the words "more force and vivacity".

Thus, the explanatory power that Hume clearly affords to both the impression-idea and simple-complex distinctions (among others) renders the Nominalist Interpretation of these distinctions untenable.[26] What is the alternative, then? That is precisely the question that I hope to answer over the next four chapters, beginning in Chapter 2 with an investigation into Hume's most explicit statements concerning the methodology of the science of human nature, continuing through discussions in Chapter 3 of Hume's understanding of substance, language, and reason, in Chapter 4 of his treatment of the science of body, and finally in Chapter 5 of Hume on necessary connection that will clear the final obstacle to presenting the full-dress answer. To put it succinctly, my proposal will be that Hume is a Nominalist about the general terms of the language of the vulgar and false philosophy, but a scientific realist about those of the scientist of human nature. Understanding the details of that two-part thesis, though, is the important and difficult business of the next four chapters to which I will now turn.

Notes

1. The following section draws from material previously presented in Landy, "Hume's Theory of Mental Representation," and Landy, "Recent Scholarship on Hume's Theory of Mental Representation."
2. By both those who endorse a very close link between the Copy Principle and Hume's theory of mental representation—Cohon and Owen, "Hume on Representation"—and those who take this link to be more distant—Garrett, "Hume's Naturalistic Theory of Representation."
3. See Landy, "Hume's Impression-Idea Distinction." The degree of force and vivacity of a perception does not affect how the perception "looks" to the mind's eye. A less forceful and vivacious image is not *faded* like an old painting. Rather, we ought to shift our focus from the idiom of "vivacity" to that of "forcefulness" and construe all of this talk functionally. A forceful idea is one that *forces* itself on the mind, which cannot be easily ignored, etc. This is of enormous help in understanding Hume's claim that any change in a perception *other than* in degree of force and vivacity change the content of the perception: if changes in force and vivacity changed the intrinsic qualities of the perception, it becomes very difficult to articulate why such changes do not also change its content. In

general, the intrinsic features of any given perception are roughly just what we normally take to be those features delivered by the various sense modalities, the perception's *imagistic* features: its particular sight, sound, smell, touch, or taste.

4. Of course, the written word "dog" does have *some* pictorial character (there is something that it is to *look like* that written word), and so it follows that an idea of this word *is* a representation of *the word*. This must be carefully distinguished, however, from what the word itself represents.
5. There is an argument to be made that Hume already includes representational content, if not as Representation$_{CO}$, then in some other form, as part of the Copy Principle. For example, Hume's initial formulation of the Copy Principle—"all our simple ideas in their first appearance are deriv'd from simple impressions, which are correspondent to them *and which they exactly represent*" (T 1.1.1.7; SBN 4, emphasis mine)—includes in its final clause an explicitly representational idiom. While it may well be that Hume ran these two theses together, it will nonetheless be worth our keeping them distinct, at the very least because they require different kinds of support, and have importantly different philosophical consequences.

   On the other hand, there are those that argue that Hume does not have a theory of representational content at all, e.g., Kail, *Projection and Realism*, 32. Kail cites Garrett's defense of the Copy Principle in Garrett, *Cognition and Commitment* as a well-established empirical generalization, not an *a priori* principle. I address Garrett on this point in Landy, "Hume's Theory of Mental Representation," using considerations similar to those just presented. The Copy Principle may well be a well-established empirical generalization, but we must distinguish this principle from the *Representational* Copy Principle, which I argue is discovered via an analysis of the idea of representation itself.
6. This way of putting the claim is too strong. *Some* complex ideas are copies of some complex impression. The more accurate claim is that only simple ideas *must* be copies. (My thanks to Hsueh Qu for pointing this out.) Still, since Representation$_{CO}$ states that only copies represent, and all complex ideas represent, even when they are not copies, cannot be Representation$_{CO}$ right.
7. Nb. by "contiguity" here, I do not mean our disposition to recall upon the experience of an object, *a*, the idea of another object, *b*, that has been experienced as contiguous with *a* in the past. Rather, what I mean is the particular associative unity that a complex spatial and/or temporal representation has in virtue of its component ideas all being contiguous with each other. My thanks to Karl Schafer for pointing out this ambiguity.
8. Of course, as Hume points out, the ideas that compose this complex idea, "are not really and in fact present to the mind, but only in power" (T 1.1.7.7; SBN 20), and so one might worry here that casting a general representation as a kind of complex idea is incompatible with Hume's commitment to the thesis that the parts that compose an actual whole must themselves actually exist. The first thing to note here is that Hume himself describes general representations as complex ideas (e.g., T 1.1.6.2; SBN 16), so there is textual support for this thesis. (More on this passage in the discussion of substance in Chapter 3.) Philosophical support is available in a number of different forms including casting the members of the revival set of a general term as the actual existing ideas that one has *at different times*, or more generally in an account of Humean dispositions that comprises as elements only actually existing perceptions, a task which is required of anyone who takes Hume to be committed to that actualist thesis.

   Hsueh Qu also raises two other objections to any account of Hume's theory of general representation that takes the content of a general representation to be at least in part constituted by the members of its revival set. First, some ideas might be part of a term's revival set at one time, but not at another, which raises the question of whether the content of the general representations correspondingly changes. Second, the ideas in one person's revival set will not be the same as those in another's, and so it appears that our general terms will have different meanings. Since Hume's account of the idea of substance goes some way towards addressing the first of these objections, I postpone responding to it until discussing the nature of that idea in Chapter 3. As the second objection is not unique to Hume, but is one with which many Modern and contemporary theories of mental content have had to grapple, its solution is outside the scope of the current study. Suffice it to say that the objection is an important one, and that I take the most plausible solution to it to lead from Hume to Kant to Hegel.

9. At least in light of Hume's thesis that, "every thing in nature is individual" (T 1.1.7.6; SBN 19). More on this in the following section.
10. What it is for a relation to be *real* in the appropriate sense, and the corresponding question of the specific form of nominalism at issue here, will be the subject of the following section.
11. Baxter, "Abstraction, Inseparability, and Identity." I present arguments against specific Ontological Interpretations, those of Tienson, Vision, and Hakkarainen, in Landy, "A Puzzle about Hume's Theory of General Representation."
12. Russow, "Simple Ideas and Resemblance," brings out why Hume must take this position and the trouble that it causes him to do so. Gamboa, "Hume on Resemblance, Relevance and Representation," presents compelling evidence that Hume accepted something like the Nominalist Interpretation, and defends Hume against objections stemming from this. Hawkins, "Simplicity, Resemblance and Contrariety in Hume's *Treatise*," is an early presentation of some of the pressing problems with the Ontological Interpretation. Nelson and Landy, "Qualities and Simple Ideas: Hume and His Debt to Berkeley," also explores the Nominalist Interpretation. For a defense of the Ontological Interpretation against the Nominalist Interpretation cf. Tienseon, "Resemblance and General Terms."
13. Berkeley, *The Works of George Berkeley, Vol. 1 of 4*, 411–12.
14. A similar point can account for the fact—raised by Vision, "Hume's Attack on Abstract Ideas: Real and Imagined,"—that while Hume's arguments address abstract general ideas, they do not address abstract particular ideas.
15. T 1.1.7.2; SBN 18.
16. Hsueh Qu raises the following objection. Consider two simultaneous perceptions: one of hunger and one of a patch of blue. While there is a sense in which one's phenomenology at that moment is complex, *prima facie* this complexity seems different from the complexity involved in a more standard case of a complex idea: the two perceptions *feel* distinct in a way that, for example, a complex impression of two colored patches next to each other do not. Notice that one can acknowledge this *feeling* of distinctness while also holding that that feeling is itself the result of coming to be able to distinguish the simple elements (however we do so!) of our complex phenomenology. That is, we can admit that these two complex ideas *feel* different, and then wonder whether they *started out* feeling different, or only became so after internalizing certain *theoretical* distinctions. Notice also that there are intermediate cases that are more complicated, for example, the phenomenology of perceiving the sights, sounds, smells, and scratches of trying to give a cat a bath. Are these deliverances from the various sense modalities experienced primarily as simple and only later combined, or as a cacophony of mutual panic that can only be parsed upon reflection?

    More generally, whereas as some scholars see it, to get talk of complex ideas and/or revival sets up and running in the first place, one needs an account of *composition*, of what makes some mereological sums of ideas into complexes, and others not, I would want to push the opposite line: it is only in deploying Hume's theory of general ideas, and so his theory of complex ideas, that we can make sense of notions such as compositionality.

17. Now, one might object here that one is *not* meant to recall this procedure in its entirety, but only the simple idea that is its focus. But what would that be like to do? Even if one could recall, for a moment, *just* the dot without any of the other perceptions that surrounded it (which we do not, in fact, do) even this image would be part of a *temporal* complex from which it would need to be distinguished, etc.
18. This example is borrowed from Nelson and Landy, "Qualities and Simple Ideas: Hume and His Debt to Berkeley."
19. My thanks to Jonathan Cottrell for urging me to consider this passage here.
20. Qu, "Hume on Mental Transparency," argues that Hume's intended conclusion here is only that we cannot fail to apprehend the qualitative characters of our current perceptions. He also cites EHU 7.13; SBN 66 and T 2.2.6.2; SBN 366 as places where Hume repeats a version of this thesis. I find even that pared down version of the thesis less philosophically plausible than does Qu, and also that the evidence to follow that shows that Hume does not hold that the simplicity of a perception is transparent to us also implies that its qualitative character cannot be. Regarding the latter, consider that if one is necessarily aware of the blueness of any blue perception, then one would also necessarily be aware that the blueness of one's complex blue perceptions is constituted by the blueness of one's simple blue perceptions, and that there is no amount of blueness smaller than that of a *minima sensibilia*.
21. It is worth noting that what is omitted by the ellipses in this quotation, and what occurs after the final sentence cited, is that Hume gives examples of indivisible perceptions—the idea of the thousandth part of a grain of sand, and the impression of the spot of ink—which might appear to be the missing introspective evidence. However, as I have already argued regarding the latter, and will argue in the next chapter regarding the former, these are not examples of simple perceptions directly perceived, but rather are complex ideas that form the revival set for the term "simple idea", which itself represents more than what is introspectively available to us.
22. In Chapter 3 we will see Hume distinguish between general ideas *simpliciter* and general ideas that have a uniting principle as their foundation. The latter are what represent the *substance* of the human mind, whereas the former represent mere modes.
23. Stroud, *Hume*, 28.
24. Landy, "Hume's Impression-Idea Distinction."
25. Hsueh Qu rightly suggests that the Nominalist Interpreter can make sense of Hume's investigation here insofar as there might be some associations that themselves depend on, not some real distinction, but some other associations. For example, we might associate all humans *because* we associate all featherless animals and all bipeds. However, Hume does frame his investigation in terms of qualities and relations rather that underlying associations, and as we are about to see, the result of the investigation appears to be a real distinction between kinds of perceptions rather than an underlying arbitrary association.
26. Qu, "Hume's (Ad Hoc?) Appeal to the Calm Passions," goes some way towards arguing that Hume's commitment to the existence and cognitive role of the calm passions is also, at least in part, the conclusion of something like an inference to the best explanation, although he understands the particular mechanisms involved in such inferences very differently than I do.

2 Hume's Scientific Realism

If the results of the previous chapter are correct, then perhaps the two most important distinctions that Hume draws in the *Treatise*, the impression-idea distinction and the simple-complex distinction, can only be made sense of by supposing that Hume took these to reflect a real difference in the ontology of the mind, despite his otherwise strong nominalist commitments. What commits him to this ontology is the work to which these two distinctions are put in explaining the regularities that Hume discovers govern our perceptions. That is, both distinctions are a kind of theoretical posit made in the course of Hume's pursuit of the science of human nature. Thus, I suggest that Hume is a kind of scientific realist. The purpose of the present chapter is to elaborate on that claim, both in delineating the details of the scientific realism that I attribute to Hume, and in defending this unorthodox interpretation.

As will become clear, the interpretation is unorthodox in at least two ways. On the one hand, one of the notable movements in Hume scholarship in the recent past is the so-called New Hume approach to Hume. While the scholars that can be placed under this heading differ from one another is a variety of ways, the common theme among them is the thesis that contrary to long-held opinion, Hume in fact believes in necessary connection and all that comes with it (e.g., robust causal powers, external existence, a substantial self, etc.). Crucial to the difference between my interpretation and the New Hume one is that New Humeans hold that while we cannot have an idea with *descriptive content* corresponding to, for example, external existence, we can employ "relative ideas" which refer to such objects, even without in any way picturing them. By contrast, I will defend the thesis that Hume holds that theoretical posits are only possible via the deployment of a perceptible model, i.e., by specifying determinate ways that the represented theoretical entity both resembles and differs from some perceptible object. So, relative ideas as the New Humeans understand them are out, and with them ideas of external existence, necessary connection, and a substantial self. New Humeans and I agree that Hume is a scientific realist, but we disagree strongly on the details of his scientific realism. In this chapter I will lay the groundwork for engaging the New Humeans by defending the thesis that all theoretical-explanatory representation requires a perceptible model for Hume, but that engagement will become explicit only in the next chapter.

On the other hand, there was a trend in Hume scholarship, especially at the turn of the previous century, to understand Hume's project in the *Treatise* as aiming at nothing more and nothing less than discovering some general principles that properly describe and systematize the behaviors of the perceptions that constitute the human mind. Such a reading is naturally paired with an account of scientific explanation such as the Deductive-Nomological Model (DN) proposed by the Logical Positivists, who of course very much saw themselves as updating Hume's own project. On that account, scientific explanation consists of subsuming particular observed phenomena under inductively-established universal propositions. For example, one might explain the observation that we tend to think of dogs just after thinking of cats by appealing to the general tendency of the human mind to associate resembling ideas.

This picture of scientific explanation has long been recognized as inadequate,[1] and yet the belief that something like it is at work in Hume has persisted. As should be clear, I aim to show that Hume has a more robust conception of scientific explanation than this, and is instead a scientific realist: an essential aspect of Hume's pursuit of that science is positing certain theoretical-explanatory entities, and he understands these posits as ontologically committing, i.e., as representing the "essence, powers, and natures" of the human mind. To begin to do so, I will engage the recent work of Graciela De Pierris, who has argued for reading Hume as what she calls an inductivist across a series of papers and in her recent book. De Pierris takes Hume to follow Newton in holding that the ultimate aim of science is to seek "assurance concerning objects, which are removed from the present testimony of our memory and senses" (EHU 7.29), and its method therefore to consist in the subsumption of observable particulars under inductively-established universal generalizations. As De Pierris puts it,

> The central idea of the Newtonian inductive method, as summarized in his Rules, is that exceptionless or nearly exceptionless universal laws are inductively derived from 'manifest qualities' of observed 'Phaenomena,' and only further observed phenomena can lead us to revise these laws.[2]

Notice that according to this characterization, Newton, and by extension Hume, understands science as concerned only, or at the very least primarily, with *manifest* phenomena, or what is observable. What I will argue here is that *contra* De Pierris, Hume's understanding of science comprises not only what is observable, but also what is in-principle unobservable as well. In particular, while De Pierris does allow for "conjecture and hypothesis" insofar as these are required to make the scope of our inductive generalizations as wide as possible, she also holds that these do not amount to inductive proofs until further observations of regular and uniform constant conjunctions are found.[3] That is, while De Pierris's Hume countenances positing unobserved phenomena as a means of widening the scope of our inductive generalizations, such phenomena are accorded a tentative epistemological status until such time as they can themselves be confirmed via observation. What I will argue is that Hume also allows for a more robust kind of theoretical activity wherein posited entities need not be observed (or observable) but can nonetheless be known to exist on the grounds that they provide an *explanation* of some manifest phenomena in need of explaining.

Specifically, I will show that while De Pierris successfully shows that, (a) Hume accepts induction as a legitimate form of scientific reasoning, (b) Hume rejects Locke's appeal to *primary qualities* as having any explanatory force, and (c) Hume rejects Locke's commitment to the *a priori* ideal of scientific explanation, this is not sufficient to show that Hume is an inductivist or that he countenances only observable phenomena. What I will argue is that De Pierris's (a)-(c) represent important conditions that Hume places on the legitimacy of theoretical posit, their conjunction leaves open the possibility that whether or not a posited entity is in-principle observable, Hume holds that whether we ought to believe in its existence depends only on its explanatory success. Regarding (a), I will show that while De Pierris is correct that Hume endorses the legitimacy of induction, establishing empirical generalizations is not the *end* of science, but rather what first *initiates* its explanatory activities. That is, it is inductively-established empirical generalizations that themselves stand in need of explanation. For example, it is because we infer the lawful correlation of a gas's temperature, pressure, and volume from observation that we find the need to posit the existence of molecules to explain these constant conjunctions. The inductively-established empirical generalization, for example Boyle's Law, is the beginning, not the end of scientific activity here.[4]

Regarding (b), another condition that Hume puts on legitimate theoretical representations is that they be formed by specifying determinate similarities and differences between that which is posited and some observable phenomenon—what I call a *perceptible model*. For example, the Bohr model of the atom explains the spectral line emissions of hydrogen by modeling the structure of the atom on that of the solar system (e.g., small particles orbiting a larger central body) while also specifying determinate differences between these (e.g., held together by electrostatic rather than gravitational forces). Importantly for Hume, requiring this sort of model disallows the use of mere *via negativa* intended to explain some observable phenomena, but doing so in name alone. It is *that* condition that underlies Hume's rejection of Locke's appeal to primary qualities, not a rejection of appeals to unobservable theoretical-explanatory entities *per se*.[5]

Finally, with respect to (c), De Pierris's thesis is that Hume rejects Locke's commitment to the *a priori* ideal of science, the notion that the ideal of scientific knowledge would be to apprehend causal powers of individual substances, "in the same way that all the properties of a triangle are contained in the idea of this figure."[6] Again, though, while De Pierris is right that Hume rejects this Lockean conception of ideal explanation, there is no reason why that rejection must be paired with the severe limitation of the scope of scientific knowledge or "proof" to manifest phenomena on which De Pierris insists. Having this sort of demonstrable intuition of the essence of objects would be *one* way to come by knowledge of their nature, but I will argue that in rejecting this ideal, Hume endorses a different means to the same end. We posit that the objects of scientific inquiry have a certain nature as a means of *explaining* certain manifest phenomena, and the explanatory success of such posits justifies our taking those objects have that nature.[7] Since the picture of scientific knowledge just sketched follows from a combination of *rejecting* the inductivist interpretation that De Pierris offers even while accepting (a)-(c), I conclude that De Pierris's evidence for (a)-(c) is insufficient to demonstrate that Hume accepts anything like here inductivist conception of scientific knowledge.

Now recall what we saw in the previous chapter of Hume's methodology in positing a real difference between impressions and ideas and between simple perceptions and complex ones. The simple-complex distinction is aimed at explaining the novelty and systematicity of human thought by appealing to the nature of perceptions themselves: that our complex phenomenology is *composed of* simple perceptions united according to certain associative tendencies. That is, it appeals to the nature of the mind to explain the inductively-established universal regularities of experience. Furthermore, as we saw, he does so via the use of a perceptible model: he specifies determinate ways in which posited simple ideas resemble and differ from complex ideas, which are perceived directly in experience. The introduction of the impression-idea distinction follows the same path. It is similarly aimed at explaining certain empirically discovered regularities, including the same variety of human thought, but also the typical differences in degrees of force and vivacity of our various perceptions. It too explains these regularities by appealing to the underlying nature of the perceptions at hand: that impressions are mental originals and ideas copies of these, etc. Farther along in this chapter, and in the ones that follow, we will see many more examples from the *Treatise* that exhibit this same structure. What such examples collectively show, especially when combined with the more holistic evidence that I will present in the following chapters, is that we need not interpret Hume as an inductivist or as accepting the inductivist model of scientific explanation. Rather, there is ample textual and philosophical support for reading Hume as a scientific realist of the kind that I will describe.

Back to De Pierris and Hume: one way to (a) accept induction, (b) reject the *a priori* ideal of scientific explanation, and (c) reject Locke's appeal to primary qualities is to accept a picture of scientific methodology and scientific explanation like the one just presented. Since it is inductively-established empirical generalizations that stand in need of explanation, induction must be valid.[8] As the *a priori* ideal of scientific explanation is the commitment to accepting purportedly *a priori* conclusions about the nature of the world in the face of countervailing empirical evidence, this too is rejected by the above account, as it is precisely *a posteriori* discoveries that demand explanation in the first place. Finally, as Locke's appeal to primary qualities combines this *a priori* ideal with the absence of any model through which to understand his purportedly explanatory posit, it too must be rejected. All of that, however, follows from a *rejection* of DN, not its acceptance. Thus, De Pierris's evidence for (a)-(c) is insufficient to demonstrate that Hume accepts anything like her inductivist conception of scientific knowledge.

My procedure here will be as follows. In the first section, I consider and critique De Pierris's interpretation of Hume.[9] In the second section, I present Hume's most explicit statements on the scope of theoretical activity to demonstrate that as Hume conceives his own project it extends well past what is "directly observable," to also include that which is "founded on the authority" of experience as well. That is, Hume understands scientific explanation as legitimately employing theoretical-explanatory posits.[10] In the third section, I examine several brief examples, in addition to those of the impression-idea distinction and the simple-complex distinction, of Hume's use of such posits to show that rather than reject their use entirely, Hume requires that such posits contain a mixture of reason and impressions, i.e., that they are validated via inferences to the best explanation, and are represented via perceptible models.

By the end of this chapter, what I hope to have shown is that Hume's methodology in the *Treatise* is sophisticatedly theoretical and explanatory, and can be captured by the following three theses that form the core of my interpretation.

THEORETICAL EXPLANATION: The form that Hume's theoretical explanations take is to explain the empirical generalities that he discovers govern the human mind by appeal to the "nature and powers" of certain theoretical posits.[11]

PERCEPTIBLE MODELS: Such posits are represented via the deployment of perceptible models, which are analogical extensions of directly observable perceptions.[12]

EXPLANATORILY ULTIMATE PRINCIPLES: While perceptible models can be used to posit theoretical entities that explain empirically discovered generalities, at a certain point such posits must themselves be taken as explanatorily basic; i.e., such posits cannot themselves be explained using the representational resources afforded by experience, and we have no other representational resources on which to draw.

De Pierris's Newtonian Inductivist Hume

Again, De Pierris argues that Hume allies himself with Newton, against Locke, in endorsing a blanket prohibition on the kind of robust theoretical-explanatory posit that I believe that Hume not only permits, but regularly employs. De Pierris takes Hume to follow Newton in believing that the end of scientific inquiry is the subsumption of observed particulars under inductively derived empirical generalizations.

> Hume's notion of inductive proof, which is at the heart of his conception of causation and scientific methodology, consists in a universalization [. . .] of our past and present uniform experience, with the attendant assumption that nature is, in Newton's words, "ever consonant with itself."[13]

De Pierris proposes that we should understand Hume's account of scientific methodology as consisting entirely in this notion of inductive proof (which she argues Hume takes over from Newton). That is, she argues that for Hume scientific practice consists in "universalization": subsuming particulars under "exceptionless universal" empirical generalizations. Note that De Pierris deliberately limits the scope of such generalizations to "our past and present uniform experience," or as we saw earlier to certain, "'manifest qualities' of observed 'Phaenomena,'."[14] This contrasts with my own proposal for understanding how Hume conceives of scientific practice in several respects, but most straightforwardly insofar as this restriction in scope implies that science is a purely descriptive enterprise that leaves no place for theoretical activity other than in anticipating what is not currently but must eventually become observable. That is, it implies that the ultimate legitimacy of belief in the objects of theoretical-explanatory posits, or "hypotheses and conjectures," depends on their being observed, and that an appeal to their explanatory success can at most accrue to them a tentatively positive epistemic status. De Pierris rejects such a conception of theoretical-explanatory posits as a live possibility for Newton and Hume by way of Rules III and IV from Newton's *Principia*, which she understands as aimed not only at rationalist advocates of the mechanical philosophy, but also Locke's version of that theory.

> Nonetheless, for both the rationalist mechanical philosophers and for Locke, the ultimate causal explanations of what we observe reside in precisely this hypothetical hidden microstructure. By contrast, Newton, as we have seen, is especially concerned that the favored hypothetical causal explanations of the mechanical philosophy do not interfere with his use of the inductive method.[15]

In fact, this passage undersells the strength of De Pierris's view. It is not just that Newton demands that theoretical-explanatory posits be *compatible* with the results of the inductive method, but also that such empirical generalizations are the only kind of explanation that can accrue genuine evidentiary force.

> Generalizations grounded by this method have the "highest evidence that a proposition can have in this [experimental] philosophy," with which no corpuscularean hypothesis or conjecture can possibly compete.[16]

Of course, it is precisely this conception of evidence that leads to De Pierris's' impoverished conception of the theoretical activity of science and eventually radical skepticism. If the evidentiary force of a scientific hypothesis is exhausted by its subsuming the observed phenomena under empirical generalizations, then it is easy enough to be skeptical of such hypotheses on the grounds that they are mere speculations, the justification for which extends only as far as the observed phenomena. That which extends beyond what is directly observed, relying as it does merely on the principle of the uniformity of nature, remains unsupported. Thus, we find De Pierris also arguing for an interpretation of Hume as a radical skeptic.

> Hume's version of the traditional theory of ideas thereby leads to a radical skeptical outcome: a quasi-perceptual form of inspection of items directly present before the mind is, from this peculiarly philosophical standpoint, the ultimate standard of justification of both our a priori and our a posteriori methods; beliefs that go beyond what is directly inspectable before the mind are then subject to skeptical attack.[17]

Because introspection provides the ultimate justification for all of our reasoning, any hypothesis that moves *beyond* introspection is thereby subject to skeptical doubt, whatever its explanatory force may be. As we will see, though, while there may be a great many subjects with regard to which Hume is a kind of skeptic, the theoretical-explanatory hypotheses of the science of human nature is definitely *not* among these.[18]

The question, then, is what leads De Pierris astray here. To see the answer to this question, we can begin by noting that De Pierris presents five aspects of Locke's conception of explanation that Newton rejects:

(1) "Locke is an advocate of the mechanical philosophy."
(2) Locke holds, "that any proper causal explanation of the operations and qualities we observe in bodies reduces to a hidden configuration of the primary qualities of their "insensible Parts.""
(3) Locke understands this microstructure by way of his account of primary qualities.
(4) "Locke retains the a priori ideal of knowledge of nature."
(5) "Locke does not anticipate an experimental method leading to the formulation of inductively established, exceptionless universal laws"[19]

De Pierris goes on to argue that Hume takes over Newton's rejection of Locke in its entirety. Her argument has two main parts: showing that Hume rejects Locke's account of primary qualities, and that Hume accepts the principle of the uniformity of nature (and is thus free to accept Newton's claims to have at least provisionally discovered certain exceptionless empirical generalizations). Establishing those two theses, however, eliminates only 3 and 5 above. Granting that Hume also rejects 4—the *a priori* ideal of knowledge of nature—and putting aside the details of Hume's rejection of the mechanical philosophy, nothing in De Pierris's argument addresses 2, the picture of scientific methodology as proceeding via appeals to the hidden microstructure of the insensible parts of observable phenomena. That is, while De Pierris is certainly right that Hume rejects Locke's notion of primary qualities, and that he accepts the legitimacy the practice of induction, establishing those two theses is not sufficient to establish further that Hume holds the impoverished conception of scientific practice that De Pierris attributes to him by way of Newton. That is, one can accept both of these theses and still hold that Hume agrees with Locke, "that any proper causal explanation of the operations and qualities we observe [. . .] reduces to a hidden configuration of [. . .] their insensible Parts," so long as these insensible parts are not construed as Lockean primary qualities. For example, I take Hume to reject Locke's conception of primary qualities for the reasons that he presents in 1.1.7 and 1.4.4, but doing so does not commit him to the rejection of theoretical-explanatory posits *simpliciter* because he is free to hold that these must always be represented via what I have been calling a *perceptible model*. To deploy a perceptible model is to form a theoretical-explanatory representation by specifying the determinate ways that the posited object resembles and differs from a model manifest phenomenon. The admittedly-hackneyed example I gave earlier was the Bohr model of the atom, which attempts to explain the spectral line emissions of hydrogen by modeling the structure of the atom on that of the solar system by specifying both determinate similarities between these (e.g., small particles orbiting a larger central body) and also differences between them (e.g., held together by electrostatic rather than gravitational forces). This is precisely the condition that Locke's account violates in positing substance as something that is *in no way like* any observed phenomena. In the following sections I will present evidence that Hume in fact holds such a view, but the point here is only that, whatever the force of that evidence, for all that De Pierris has argued, Hume *could* still hold such a view.

Furthermore, Hume's acceptance of certain inductively-discovered empirical generalizations alone does not commit him to an inductivist view of scientific practice (according to which it ultimately consists *entirely* of discovering such generalizations). Notice, for example, that on the kind of account that I presented earlier, the *explicanda* of a scientific theory is exactly that the observed particulars obey the empirical generalizations that they do, a phenomenon that would be impossible to encounter were induction not legitimate. What De Pierris's Newton rejects is the claim that *a priori* reasoning can legitimately be employed in justifying the use of a theoretical-explanatory posit. Hume can accept that claim without having to accept the further claim that the only way to justify such a posit is by eventually observing it. For example, one can reject the *a priori* method for formulating such hypotheses, and also hold that it is a theoretical posit's success in explaining some otherwise surprising or puzzling manifest phenomenon that justifies our believing in its existence. For example, one might justify a belief in the particulate theory of gases on the basis of its success in explaining the Ideal Gas Law (which is itself an inductively-established universal generalization).

Finally, De Pierris presents texts that she takes to show that Hume endorses Newton's rejection of theoretical-explanatory posits in their entirety, for example, the following passage from the *Enquiry*.

> It was never the meaning of Sir ISAAC NEWTON to rob second causes of all force or energy; though some of his followers have endeavoured to establish that theory upon his authority. On the contrary, that great philosopher had recourse to an etherial active fluid to explain his universal attraction; though he was so cautious and modest as to allow, that it was a mere hypothesis, not to be insisted on, without more experiments.
>
> EHU 7.25 n16; SBN 73 n1

As De Pierris notes, Hume is here referring to Newton's attempt in Query 21 of the *Opticks* to explain the law of universal gravitation by appealing to the pressure exerted by the differing densities of the Aether produced by massive bodies. De Pierris takes Hume's description of Newton's conjecture as "mere hypothesis" to indicate Hume's agreement with Newton that while such explanatory posits might be tried, their ultimate justification can come only by way of direct observation. Notice, though, that Hume does not claim that the existence of ether needs to be *directly observed* to be proved, but only that "more experiments" are necessary to establish it. Such experiments might yield a direct observation of the ether, but that is not the only kind of experimental evidence that can speak in favor of that hypothesis. For example, one way that a theoretical posit might be proved is by collecting further instances of otherwise surprising phenomena that the posit can explain and that other competing hypotheses cannot. For example, its ability to successfully explain a wide variety of experimental results that no other theory can is the most significant piece of evidence in favor of the so-called Standard Model of contemporary particle physics, despite the fact that the objects that it posits cannot be directly observed. Merely assigning the ether hypothesis a tentative status does not commit Hume to assigning that same status to all theoretical-explanatory posits *per se*.

Notice also, that Hume refers to Newton's conjecture as an attempt "to *explain* his universal attraction." The law of universal gravitation is an empirical generalization that holds of all observed particulars (massive bodies). If the aim of science were merely to subsume manifest phenomena under universal generalizations, then there would be no demand to *explain* such discoveries: scientific inquiry would end with their formulation, and no such explanation would be necessary.[20] Hume acknowledges, however, that the law of universal gravitation does call for an explanation in terms of the nature of the underlying substances at hand. Newton's conjecture does just this: it explains the inductively derived generalizations by positing an unobserved fluid that resembles and differs from the observed phenomenon on which it is modeled in determinate ways. Thus, insofar as Hume approves of the *form* of Newton's conjecture, if not it's then current evidentiary force, it is because he is a scientific realist, not an inductivist.

If Hume does approve of the form of Newton's conjecture, though, what then of his idiom of "mere hypothesis"? The first thing to note about that phrase is that it is not one that Hume explicitly endorses, but rather it occurs in his description of how Newton himself understood his conjecture. The second thing to note about it is that Hume takes Newton to have understood his conjecture in this way because "he was so cautious and modest," and as Schliesser has noted, it is not clear that Hume's descriptions of Newton as cautious and modest are meant to be flattering.[21] Hume employs the same idiom in his description of Newton in *The History of England* to deliver a series of subtle backhanded compliments.

> In Newton this island may boast of having produced the greatest and rarest genius that ever arose for the ornament and instruction of the species. From modesty, ignorant of his superiority above the rest of mankind; and thence, less careful to accommodate his reasonings to common apprehensions: More anxious to merit than to acquire fame: He was, from these causes, long unknown to the world; but his reputation at last broke out with a lustre, which scarcely any writer, during his own life-time, had ever before attained.
>
> H VI, 542

While Hume is generally laudatory here, he does also cite Newton's modesty as the source his misconstruing, and more specifically his *underestimating*, the credit that is due to him (and of his failure to properly disseminate his work). It is easy enough to think of Hume as holding a similar position with respect to Newton's modesty regarding the explanatory force of his account of gravity, i.e., of taking Newton's epistemic modesty to result in his misconstruing the nature of his hypothesis and underestimating its explanatory force.[22]

There is another passage form the *Enquiry* in which Hume discusses gravity and its ultimate cause that also appears to support De Pierris's interpretation of Hume as holding that the end of scientific inquiry is merely subsuming particular manifest phenomena under inductively-established universal generalizations.[23]

> It is confessed, that the utmost effort of human reason is, to reduce the principles, productive of natural phaenomena, to a greater simplicity, and to resolve the many particular effects into a few general causes, by means of reasonings from analogy, experience, and observation.

> But as to the causes of these general causes, we should in vain attempt their discovery; nor shall we ever be able to satisfy ourselves, by any particular explication of them. These ultimate springs and principles are totally shut up from human curiosity and enquiry. Elasticity, gravity, cohesion of parts, communication of motion by impulse; these are probably the ultimate causes and principles which we shall ever discover in nature; and we may esteem ourselves sufficiently happy, if, by accurate enquiry and reasoning, we can trace up the particular phaenomena to, or near to, these general principles. The most perfect philosophy of the natural kind only staves off our ignorance a little longer.
>
> EHU 4.12; SBN 30–1

Hume appears to sketch precisely the picture of scientific inquiry that De Pierris attributes to him. The end of scientific inquiry is formulating maximally simple and general principles derived from experience. We cannot so much as investigate, much less gain knowledge of the causes underlying these principles. Even the most perfect science leaves us only a little less ignorant of such ultimate causes. That looks pretty bleak for anyone who holds that Hume is no mere inductivist.

A few preliminary points about this passage. First, the context in which it appears is important. Here is the passage preceding this one, in which Hume argues that causes cannot be discovered *a priori*.

> In a word, then, every effect is a distinct event from its cause. It could not, therefore, be discovered in the cause, and the first invention or conception of it, *a priori*, must be entirely arbitrary. And even after it is suggested, the conjunction of it with the cause must appear equally arbitrary; since there are always many other effects, which, to reason, must seem fully as consistent and natural. In vain, therefore, should we pretend to determine any single event, or infer any cause or effect, without the assistance of observation and experience.
>
> EHU 4.11; SBN 30

This passage provides important context in at least two ways. First, it makes it tenable that Hume's endorsement of such a radically empiricist position in the later passage may well be a hyperbolic reaction against what he saw as an extremist view on the other side of the debate (held by the great Mr. Locke, no less!). In railing against the *a priori* ideal of scientific knowledge, he lets the pendulum swing too far in the opposite direction for a moment.[24] This supposition is supported by the second way that this passage provides context. Notice that Hume's conclusion here is only that inferring any cause or effect requires observation and experience. As we have noted previously, that is a thesis that the scientific realist and the inductivist can both endorse, and so is not evidence in favor of either interpretation.

Immediately following this conclusion, however, Hume opens the later passage thusly:

> *Hence we may discover* the reason, why no philosopher, who is rational and modest, has ever pretended to assign the ultimate cause of any natural operation, or to show distinctly the action of that power, which produces any single effect in the universe.
>
> EHU 4.12; SBN 30, emphasis mine

Hume apparently takes what he says in the later passage to *follow from* his conclusion in the earlier one. As we have just seen, though, the radically impoverished empiricism of the later passage does not follow from the conclusion of the earlier one because that conclusion is compatible with a more robust scientific realism. Thus, we have a choice: we can read Hume as making an invalid argument from the claim that we cannot know causes *a priori* to the conclusion that science aims only at universal generalizations, or we can read him as exaggerating the role of experience in the face of what he takes to be the extremist view that we can discover causal relations, "by the mere operation of our reason, without experience" (EHU 4.8; SBN 28). I would suggest that the latter is at the very least a viable option.[25]

Thus, while De Pierris makes a compelling case that Hume follows Newton in rejecting certain of Locke's theses regarding scientific methodology, I find that the evidence that she presents that Hume also adopts what she takes to be Newton's inductivism insufficient. As that is a deliberately impoverished understanding of scientific activity that has radically skeptical consequences, we should require a great deal of evidence before attributing it to a philosopher of Hume's stature, especially since as we are about to see Hume appears to rely on a very different understanding of science in his own pursuit of the science of human nature, and since his most explicit pronouncements of the subject support this practice. Hume agrees with Newton on the importance of empirical generalizations for the explanatory work of science, but also allows for the deployment of genuinely *explanatory* hypotheses, which not only describe both the observed and as of yet unobserved phenomena, but also explain these by appeal to their underlying, sometimes unobservable, nature. Where Hume differs from Locke is in demanding that such theoretical posits be made intelligible via being modeled on some more familiar observable phenomenon. Thus, De Pierris is correct to cite, "the postulation of a hidden microstructure of primary qualities or properties of bodies," as something that Hume finds objectionable in Locke, but wrong to think that what is objectionable here is the *postulation* of something hidden rather than the fact that the idea of *what* Locke takes to be hidden, "primary qualities", has no content.

Thus, I conclude that the case for reading Hume as an inductivist about scientific methodology is insufficient, and that given the philosophic implausibility of that account, we ought to seek to understand Hume as having a more plausible approach to his own scientific methodology.[26]

The Scope of Theoretical Activity

As I mentioned above, this interpretation of Hume is different from many of the most prominent extant interpretations, and it will therefore be necessary to marshal even more textual evidence in support of it. Much of this support will be holistic and will come in the form of its success in accounting for Hume's actual pursuit of the science of human nature throughout Book I. (I will turn to the structure of that engagement in the next chapter.) Still, there are moments throughout Book I at which Hume briefly turns his attention to the meta-philosophical issues in this domain, and it will be worthwhile to take the time to consider these texts before proceeding.

Hume is perhaps most explicit about his vision of the science of human nature in the Introduction and while scholars seem to find support there for reading Hume as engaging in a purely descriptive project,[27] a closer consideration reveals that Hume's statements there all include important caveats that temper his apparent contempt for theoretical-explanatory activity. What I will do here, then, is walk through the passages in the Introduction where Hume's attention is focused on describing the methodology and anticipated results of the science of human nature (roughly Intro.7–10; SBN xvi-xix), paying careful attention to the ontological commitments that Hume does and does not undertake and reject. We can begin with Hume's comparison of the epistemological status of science of human nature to that of the science of external bodies.

> For to me it seems evident, that the essence of the human mind being equally unknown to us with that of external bodies, it must be equally impossible to form any notion of its powers and qualities *otherwise than from careful and exact experiments, and the observation of those particular effects, which result from its different circumstances*.
>
> T Intro.8; SBN xvii, emphasis mine

Far from claiming that we cannot know the "powers and qualities" or the human mind, Hume is in fact presenting the precise methods that the science of human nature will use to discover these. It is through "careful and exact experiments" and "the observation of those particular effects, which result from its different circumstances" that we *do* come to know the essence of the human mind. Of course, Hume does start out this sentence by declaring that the essence of the human mind is unknown to us (or, more precisely, *as unknown as* that of external bodies), but he is there making a claim about our state of knowledge *before* we make such experiments and observations. There is no other way to make sense of his methodological recommendation, the end of which is to remedy this ignorance. Thus, this passage is a clear statement of the realist aims of science.[28]

Additionally, notice that there is nothing here with which a scientific realist need disagree. Experimentation and observation of the behavior of the objects of study in various circumstances are precisely what provide the foundation for our theoretical-explanatory activities. What is discovered by such activities, when properly conducted, is the essence, the real powers and qualities, of the object studied. Again, far from being a condemnation of theoretical speculation, this passage is a guide to its proper implementation and an endorsement of its realist aims. This should make us wonder, though, to whom Hume takes himself to be aiming this recommendation, and the answer to that question is easy enough to find given that Hume credits the pioneers of this method as "some late philosophers in *England*" (T. Intro 7; SBN xvii), and specifically cites Locke, Shaftesbury, Mandeville, Hutchenson, and Butler. That list certainly includes philosophers with a much more robustly realist approach to scientific explanation than Hume has been taken to have, and contrasts most directly with their Continental "rationalist" counterparts. That is, when Hume recommends experimentation and observation, he is not warning against scientific realism, but only against entirely non-empirical (e.g., *a priori*) methods. That this is Hume's intended target here will become clearer as we consider the remainder of these passages.

Now consider the passage that immediately follows this one.

> And tho' we must endeavour to render all our principles as universal as possible, by tracing up our experiments to the utmost, and explaining all effects from the simplest and fewest causes, 'tis still certain we cannot go beyond experience; and any hypothesis, that pretends to discover the ultimate original qualities of human nature, ought at first to be rejected as presumptuous and chimerical.
>
> T Intro.8; SBN xvii

The opening of this passage does appear to confirm the view of Hume as an inductivist: "we must render our principles as *universal* as possible" resonates with De Pierris's interpretation according to which scientific methodology consists in subsuming particulars under universal empirical generalizations, and "'tis certain that we cannot go beyond experience," sounds like her claim that what is most real are observed phenomena, not the theoretical posits that explain these. Of course, the first clause is offset by what follows it—"explaining all effects from the simplest and fewest causes"—which makes explicit appeal to specifically *causal* explanations, which sits less well with the inductivist picture. And the second is followed by what appears to be an explication of the proscription against going beyond experience, which limits its scope to "any hypothesis that pretends to discover the ultimate original qualities of human nature." To understand this more limited version of the prohibition, we must first properly understand what Hume means by that phrase. What Hume is proscribing is the attempt to move beyond what the science of human nature must take as explanatorily basic. That is, in explaining the features of the human mind, the scientist of human nature will posit certain entities and principles. To have the representational content required to fulfill their explanatory function, these posits must be modeled on directly observable phenomena, which for Hume must be perceptions themselves. One might inquire into the explanation of such entities or principles, and if there is a further perceptible model available to answer that inquiry, then a legitimate explanation might be given. At some point, however, such resources are exhausted and no further legitimate explanation is available. Hume believes that at that point philosophers are all too often tempted to make use of *illegitimate* explanatory hypotheses—e.g., ones that do not employ perceptible models, and so are devoid of any real content—and this is what he is here recommending against: "any hypothesis, that pretends to discover the ultimate original qualities of human nature." This forbearance is not against explanatory hypotheses *in general*, but only against those that pretend to discover ultimate original principles where only "provisional" ones are legitimately possible.

Hume reprimands the philosopher that pursues such explanations in the following paragraph.

> I do not think a philosopher, who wou'd apply himself so earnestly to the explaining the ultimate principles of the soul wou'd show himself a great master in that very science of human nature, which he presents to explain, or very knowing in what is naturally satisfactory to the mind of man.
>
> T Intro.9; SBN xvii-xviii

It is not the philosopher that makes an initial appeal to ultimate explanatory principles that is Hume's target here, but rather the philosopher that aims to explain these principles themselves, which being ultimate can receive no such explanation.[29] What such a philosopher misunderstands is that the ceaseless pursuit of further and further explanations is not what is satisfactory to the mind of man, but rather that it is the termination of such explanations in something explanatorily basic that is so. That is, Hume's point here is not that we must not move beyond a mere descriptive phenomenology, but only that explanations must come to an end where our ability to ground those explanations in experience does, and that we must resist the temptation to pretend to move beyond this point with "presumptuous" and "chimerical" hypotheses.

This entire dialectic of reaching explanatory rock bottom, being tempted to appeal to unfounded first principles, and the failure of philosophers who give in to this temptation is repeated numerous times throughout Book I.

Here is an example from 1.1.4 "Of the connexion or association of ideas" in which Hume uses the same idiom of "original qualities of human nature."

> Here is a kind of ATTRACTION, which in the mental world will be found to have as extraordinary effects as in the natural, and to show itself in as many and as various forms. Its effects are every where conspicuous; but as to its causes, they are mostly unknown, and must be resolv'd into *original* qualities of human nature, which I pretend not to explain. Nothing is more requisite for a true philosopher, than to restrain the intemperate desire of searching into causes, and having establish'd any doctrine upon a sufficient number of experiments, rest contented with that, when he sees a farther examination wou'd lead him into obscure and uncertain speculations. In that case his enquiry wou'd be much better employ'd in examining the effects than the causes of his principle.
>
> T 1.1.4.6; SBN 12–13

Notice that here Hume's recommendation to the philosopher "to restrain the intemperate desire of searching into causes" is limited in its scope to "when he sees a farther examination wou'd lead him into obscure and uncertain speculations." That is, there is nothing *per se* illegitimate about searching into causes, but since the desire to do so is intemperate, one must be careful to recognize when its legitimate fulfillment has been completed. Here, for example, Hume recognizes that an attempt to explain the associations of ideas would go beyond the resources that are available to him, and so must rest content with taking these to be "*original* qualities of human nature," i.e., with taking them to be explanatorily basic, i.e., he must "pretend not to explain" them. What this series of claims suggests, though, is that the form that such an explanation *would* take, where legitimate, would be in appealing to the unobserved causes of these associations.

In fact, Hume makes it clear that this is the form that such an explanation would take when he revisits his decision to take the three associations of ideas as explanatory primitives in his discussion of the idea of vacuum, and there speculates about what (chimerical) explanations he might have given of those relations.

> When I receiv'd the relations of *resemblance*, *contiguity*, and *causation*, as principles of union among ideas, without examining into their causes, 'twas more in prosecution of my first maxim, that we must in the end rest contended with experience, than for want of something specious and plausible, which I might have display'd on that subject. 'Twou'd have been easy to have made an imaginary dissection of the brain, and have shown, why upon our conception of any idea, the animal spirits run into all the contiguous traces, and rouze up the other ideas, that are related to it.
>
> T 1.2.5.20; SBN 60

This passage appears to raise a new problem for the line that I have proposed, because it makes it clear that Hume in fact does take himself to have a perceptible model available to explain the association of ideas: a dissection of the brain. That in turn implies that it is not the unavailability of such a model that limits the scope of theoretical activity in this case, but rather that it is the *speculative* nature of employing such a model that does so. So, should we take Hume's proscription to be against theoretical activity *per se*? There are two reasons to resist this conclusion.

The first is to consider what makes Hume's appeal to an imaginary dissection of the brain both plausible but also specious. What makes it plausible is the *appearance* that the associations of ideas are being explained by something that is better understood than they are. As Hume describes, we can picture a brain being dissected and discovering certain animal spirits there that we observe to behave in ways that correspond to the associations of our ideas, etc. Such a fantasy has the air of explanation because it seems to ground the associations of ideas in the essence and powers of something distinct from them: the human brain. The reason that this explanation is specious, though, is that after 1.2.6 and 1.4.2, we know that the idea of the brain is, in fact, not an idea of external existence, but only of the apparent constancy and coherence of some of our perceptions. That constancy and coherence, though, is exactly the constancy and coherence of the associations of ideas, and so cannot be used to give an informative explanation of the latter. So, what appears to be an explanation of these associations turns out to be circular, insofar as the *explanans* will find its own ultimate and original principle in the *explicandum*.

The second reason to resist taking the speciousness of this explanation to reflect a more wide-scope failure of theoretical explanation comes from the continuation of this passage.

> But tho' I have neglected any advantage, which I might have drawn from this topic, in explaining the relations of ideas, I am afraid I must here have recourse to it, in order to account for the mistakes that arise from these relations.
>
> T 1.2.5.20; SBN 60

Whereas in his earlier consideration of the relations of ideas there was no particular demand for their explanation, in considering the phenomenon of our mistakenly taking ourselves to have an idea of vacuum such a demand does present itself. That is, as we would expect, an explanatory appeal to theoretical entities is not warranted *merely* by the availability of a perceptible model for those entities, but other conditions must also be met to make such an appeal epistemically legitimate. For one, there must be some phenomenon in need of explanation. This understanding accords with the thesis that I have been attributing to Hume that explanations must terminate, and that we must resist the illegitimate temptation to explain what is explanatorily basic. An explanation is neither necessary nor possible for every phenomenon, and where there is no specific demand for explanation, one must resist giving such explanations merely for their own sake.

Finally, as Hume mentions later, the principles of associations by themselves do not legitimize such theoretical-explanatory activity in part because they are not, in fact, *universal* generalizations.

> These principles I allow to be neither the *infallible* nor the *sole* causes of an union of ideas. They are not the infallible causes. For one may fix his attention during some time on any one object without looking farther. They are not the sole causes. For the thought that has evidently a very regular motion in running its objects, may leap from the heavens to the earth, from one end of the creation to the other, without any certain method or order. But tho' I allow this weakness in these three relations, and this irregularity in the imagination; yet I assert that the only *general* principles, which associate ideas, are resemblance, contiguity and causation.
>
> T 1.3.6.13; SBN 92–93

If perceptions unfailingly conformed to the principles of association, it might be proper to conclude that there is something about their nature or essence that required such behavior. (Or as in the case of gases conforming to the Ideal Gas Law, if perceptions conformed closely enough to such principles and the difference could itself be explained by appeal to their nature or essence.) That perceptions do not so conform preempts the need for an explanation of those principles, although as Hume notes these principles remain the best general descriptions that we can give of the regularities that perceptions do exhibit.

The associations of ideas are explanatorily basic not because explanations cannot make appeal to unobserved theoretical entities or principles, but rather because there is neither a legitimate perceptible model available to explain these, nor a particular demand to do so. In both the passage from the Introduction and in Hume's discussion of the associations of ideas, all that Hume recommends is that philosophers learn to resist the temptation to explain *everything*, and to accept that in pursuing the science of human nature, we will inevitably hit explanatory bedrock where the phenomena observed, the inductive generalizations arising from these, or the theoretical posits that themselves serve as explanations of such generalizations, simply cannot be further explained.

With this more modest interpretation in mind, we can now proceed to the next passage from the Introduction.

> *When we see, that we have arriv'd at the utmost extent of human reason*, we sit down contented; tho' we be perfectly satisfy'd in the main of our ignorance, and perceive that we can give no reason for our most general and most refin'd principles, beside our experience of their reality;

> which is the reason of the mere vulgar, and what it requir'd no study at first to have discover'd for the most particular and most extraordinary phænomenon. And as *this impossibility of making any farther progress* is enough to satisfy the reader, so the writer may derive a more delicate satisfaction from the free confession of ignorance, and from his prudence in avoiding that error, into which so many have fallen, of imposing their conjectures and hypotheses on the world for the most certain principles.
>
> T Intro.9; SBN xviii, emphases mine

Again, while at first glance it might appear troubling to the scientific realist that Hume claims that, "we can give no reason for our most general and most refin'd principles, beside our experience of their reality," it is clear enough from the context that this is not a general forbearance on theoretical-explanatory activity, but rather that that proscription is limited in its scope specifically to those who fail to recognize the "impossibility of making any farther progress" past the point at which "we have arriv'd at the utmost extent of human reason." The first sentence here makes clear that the scientist of human nature is not only allowed to employ "general and most refin'd principles" (explanatory hypotheses), but also to reach these via the employment of not just experience, but also *reason*. So, Hume's claim that we experience the reality of these principles should not be understood as a claim that the only legitimate explanations are the generalizations of descriptive phenomenology. Rather, it is that we reach the limit of our theoretical activity when we can no longer appeal to experience as its *ground*. That is, when our explanations are no longer held accountable to experience for their evidentiary support or are no longer represented via what I have been calling a perceptible model.[30] Once again, there is nothing here to which a scientific realist must object. Hume's critique is aimed at those who would pretend to explain the ultimate experience-based principles on non-experiential ones.[31]

In fact, Hume returns to the matter of employing refin'd principles in the conclusion of Book I, where he explicitly considers a general prohibition on these.

> Shall we, then, establish it for a general maxim, that no refin'd or elaborate reasoning is ever to be receiv'd? Consider well the consequences of such a principle. By this means you cut off entirely all science and philosophy: You proceed upon one singular quality of the imagination, and by a parity of reason must embrace all of them: And you expressly contradict yourself; since this maxim must be built on the preceding reasoning, which will be allow'd to be sufficiently refin'd and metaphysical. [. . .] Very refin'd reflections have little or no influence upon us; and yet we do not, and cannot establish it for a rule, that they ought not to have any influence; which implies a manifest contradiction.
>
> T 1.4.7.7; SBN 268

While there is much to be said about the conclusion of Book I more generally, for present purposes it will suffice to notice Hume's very strong language here. He concedes that his own procedure in conducting the science of human nature, "will be allow'd to be sufficiently refin'd and metaphysical," and wonders whether this implies that he should on those grounds reject it. His answer: such a forbearance would "cut off entirely all science and philosophy" and ceasing all such activity "implies a manifest contradiction." While philosophy and the science of human nature both of which employ refin'd reasoning (explanatory hypotheses) might well cause melancholy and delirium (T 1.4.7.9; SBN 269), they are also entirely necessary to correcting the beliefs of the vulgar and of the false philosophy and avoiding manifest contradictions (e.g., those that result in the ideas of necessary connection, external existence, and personal identity).[32] As Hume notes just a few pages later, the science of human nature, for all of its refin'd principles, is itself the very pinnacle of human knowledge, and we have every reason to believe that it will ultimately triumph, because of its explanatory power, over its competitors.

> But were these [specious] hypotheses once remov'd, we might hope to establish a system or set of opinions, which if not true (for that, perhaps, is too much to be hop'd for) might at least be satisfactory to the human mind, and might stand the test of the most critical examination. Nor shou'd we despair of attaining this end, because of the many chimerical systems, which have successfully arisen and decay'd away among men, wou'd we consider the shortness of the period, wherein these questions have been the subjects of enquiry and reasoning. [. . .] Human Nature is the only science of man; and yet has been hitherto the most neglected.
>
> T 1.4.7.14; SBN 272–3

Perhaps the most prominent challenge that has faced proponents of scientific realism in recent years is the pessimistic meta-induction—roughly the argument that since all previous scientific theories have been false, we should conclude on inductive grounds that all future ones will be as well—and here we find Hume anticipating and responding to this very challenge, not on the grounds that there is anything wrong with induction, but rather on those that the inductive base on which this argument would be made is too small to be considered representative! That historical tidbit aside, the point here is that Hume rejects wholeheartedly a prohibition on refined reasoning, and explicitly endorses the view that such reasoning is the only way that we will stand any chance of reaching a true and explanatory scientific theory.

All of which brings us to one last passage from the Introduction that again appears to cast Hume as an inductivist, but which is actually an expression of his scientific realism:

> None of them [all arts and sciences] can go beyond experience, or establish any principles which are not founded on that authority.
>
> T Intro.10; SBN xviii–xix

While the first half of this sentence can seem to lend support to the notion that Hume holds that the science of human nature must not include anything but descriptive phenomenology and perhaps a few defeasible empirical generalizations derived from such a description, the second half of the sentence extends the scope of the sciences to include those principles which are *founded on* that authority. Again, that is not something to which the scientific realist would disagree. Both the content of and evidence for a theoretical posit will be derived from the authority of experience in the sense that the theoretical posits of the science of human nature are legitimate just in case they explain some phenomenon using a *perceptible model*.[33] That perceptible model does "go beyond experience" insofar as it will be an analogical extension of what is found in experience itself, but Hume's disjunction here allows for precisely such extensions. To see this, we can look at two other passages from later in Book I that exhibit the same structure as this one.

Here is Hume commenting on the custom that underlies the use of general terms.

> To explain the ultimate causes of our mental actions is impossible. 'Tis sufficient, if we can give any satisfactory account of them from experience and analogy.
>
> T 1.1.7.11; SBN 22

The first sentence here again can be read as if Hume is issuing a universal prohibition on theoretical-explanatory science, but the second sentence indicates that he is doing no such thing. That we cannot explain the ultimate causes of our mental actions is due simply to the fact that they are *ultimate*. Again, at some point, something must be taken as primitive, an unexplained explainer, and Hume is sounding this cautionary note lest his reader's expectations become too ambitious. That this is the case can be discerned by the second sentence, in which Hume once again extends the scope of legitimate scientific inquiry beyond merely what is experienced to include also that which can be understood via *analogy*, i.e., what can be understood via an analogical extension of experience.[34]

Here is the second passage, this time from Hume's discussion of the idea of body in 1.4.2. Notice that Hume here contrasts what is an object of the senses with what is (legitimately) "deriv'd from experience and observation." That contrast, like the two preceding ones, depends on the scope of theoretical activity outstripping mere descriptive phenomenology.

> As to the *independency* of our perceptions on ourselves, this can never be an object of the senses; but any opinion we form concerning it, must be deriv'd from experience and observation
>
> T 1.4.2.10; SBN 191

Of course, such passages do not give any *details* of what it is to derive an opinion from experience and observation, or to give an adequate account from analogy, or to establish principles found on the authority of experience. What they do, however, is indicate a contrast between mere descriptive phenomenology and a different kind of theoretical-explanatory activity that at once moves beyond experience, while still bearing some content- and authority-dependent connection to experience. Once again, this is exactly how the scientific realist, not the instrumental or the rationalist, understands the role of theoretical activity.

Now, finally, with those apparently damning passages situated in their proper context, we can turn to a passage from the Introduction that more explicitly supports this interpretation of Hume's understanding of the methodology of the science of human nature. Here Hume proclaims that the truths at which his pursuit of this science aims cannot be either easy to observe or obvious:

> For if truth be at all within the reach of human capacity, 'tis certain it must lie very deep and abstruse; and to hope we shall arrive at it without pains, while the greatest geniuses have failed, must certainly be esteemed sufficiently vain and presumptuous. I pretend to no such advantage in the philosophy I am going to unfold, and would esteem it a strong presumption against it, were it so very easy and obvious.
>
> T Intro. 3; SBN xiv–xv

The conclusions that the scientist of human nature will reach are not ones that are "easy or obvious," but "lie very deep and abstruse." While it would certainly be no simple matter to provide a complete and accurate descriptive phenomenology of the human mind, such a study would be the precisely opposite of deep and abstruse: it would be intentionally restricted to what lies on the surface, so to speak, of that phenomenology and to what is least obscure about it. In fact, in the following paragraph Hume explicitly cites the main advantage that accrues to the scientist who begins his studies with human nature as being his or her ability to "*explain the nature* of the ideas we employ" (T Intro. 4; SBN xv). It is not that the scientist of human nature perfectly describes those ideas that make that science the most fundamental, but rather that he or she is able to *explain* their *nature*.

Of course, the Introduction to the *Treatise* is not the only place that Hume addresses the nature of science, but careful attention to other texts yields similar results: if one begins by thinking that Hume is an inductivist, these texts appear to confirm that view, but closer examination belies the realist Hume. Consider, for example, this passage from the *Enquiry*.

> For surely, if there be any relation among objects, which it imports to us to know perfectly, it is that of cause and effect. On this are founded all our reasonings concerning matter of fact or existence. By means of it alone we attain any assurance concerning objects, which are removed from the present testimony of our memory and senses. The

> only immediate utility of all sciences, is to teach us, how to control and regulate future events by their causes.
>
> EHU 7.29; SBN 76

If one is inclined to understand Hume as an inductivist, the final sentence here screams a confirmation of that interpretation: the point of science is "to control and regulate future events," i.e., science aims at neither truth nor explanation, and does not purport to represent anything other than observable phenomena. What Hume actually says, though, is that such control is "the only immediate *utility*" of science. That is, what science is *used for* is to control future events. That does not imply that science does not *also* aim at discovering the true ontology of the world or the underlying explanations of observable phenomena. Rather, as Hume famously notes in the Conclusion of Book I, the *practical* benefits of science and its discovery of truth are often at cross-purposes (T 1.4.7.9; SBN 269).[35] While the *practical* utility of science may be in prediction and control, its *theoretical* uses are aimed squarely at truth and explanation. Finally, and most importantly, notice that in this passage that the claim regarding the utility of science comes as the final entry in a list of reasons why knowledge of causal relations is so important, the preceding members of which are that such knowledge underlies "all our reasonings concerning *matters of fact or existence*" (truth and ontology) and "assurance concerning objects, which are removed from the present testimony of our memory and senses" (unobservables). So even here the *practical* utility of science is only one of science's aims, and the *theoretical* benefits listed are distinctly realist, not merely inductivist.

So it seems that rather than containing any vehement denials of the legitimacy of theoretical-explanatory activities or enthusiastic commitments to a purely descriptive phenomenology, the Introduction, like many other places in the Book I, carefully hedges such pronouncements to give a very different picture of the science of human nature than has previously been supposed. What Hume commits himself to there are the following theses:

(1) In pursuing the science of human nature, something will have to be taken as explanatorily basic, and philosophers must resist pretending to explain these ultimate principles by appealing to that which is not suitably derived from experience.
(2) Theoretical-explanatory posits depend for their content and legitimacy on their derivation from experience, but this derivation can move beyond experience via reason and analogy.

These theses are not only compatible with the picture of scientific realism that I presented in the opening section, but in fact follow from it. Having examined some of Hume's general reflections on the methodology of the science of human nature, it will now be instructive to examine a few brief further examples of Hume's putting that methodology into practice in order to fill out the picture that we have just been sketching in its most broad strokes.

Representing Theoretical Entities

In the previous section we saw that far from committing himself to conducting the science of human nature purely via descriptive phenomenology, Hume's explicit statements about the methodology of that science are compatible with and even support the use of theoretical-explanatory posits so long as these are properly derived from experience. In the previous chapter, we saw this methodology play out in the cases of the impression-idea and simple-complex distinctions. In this section, I present further examples of Hume employing this methodology in the hopes of giving some further details to the picture just sketched. I aim to show that Hume himself relies on the use of theoretical-explanatory entities. Hume takes these to be derived from experience insofar as they are represented via models that are created by specifying determinate ways that the posited entity is similar to and differs from that experience on which they are modeled, and that the legitimacy of employing such a model stems from the need to explain some experiential phenomenon. Furthermore, it will be important to note that in these cases the theoretical-explanatory posits at issue are the product of a combination of experience and *reason*, rather than the imagination.[36]

To that end, I will begin with a telling passage from Hume's discussion of the idea of vacuum. An inductivist reading of Hume makes understanding this passage, as well as the others considered in this section, very difficult to understand, whereas a realist reading handles them with ease. In this first passage, Hume considers an objection to his thesis that we can form no idea of a vacuum that calls for Hume to account for what happens when we imagine a room to be emptied of whatsoever fills it, while the structure of the room remains the same. Here is Hume's description.

> When every thing is annihilated in the chamber, and the walls continue immoveable, the chamber must be conceiv'd much in the same manner as at present, when the air that fills it, is not an object of the senses. This annihilation leaves to the *eye*, that fictitious distance, which is discover'd by the different parts of the organ, that are affected, and by the degrees of light and shade; and to the *feeling*, that which consists in a sensation of motion in the hand, or other member of the body. In vain shou'd we search any farther. On which-ever side we turn this subject, we shall find that these are the only impressions such an object can produce after the suppos'd annihilation; and it has already been remark'd, that impressions can give rise to no ideas, but such as resemble them.
>
> T 1.2.5.23, SBN 62–63

This is a remarkable passage for a number of reasons. First, Hume's argument here simply takes for granted that before engaging in the objector's thought experiment, we represent the room as filled with air that "is not an object of the senses." In itself, that strongly implies that Hume holds that we can represent unobservables, and that understanding how we represent a room as filled with such air should give us some insight into this issue. Second, Hume continues by noting that whatever the suggested annihilation might otherwise accomplish, it does not change the *image* that is the representation of the room: the same degrees of light and shade remain, as do the feelings of motion in the hand, etc. So, again, what is represented by the senses in representing the room as filled with air or not is constant. "In vain shou'd we search any farther." What changes is not what is sensed, but something else, something that is not itself experienced.

Finally, notice that despite the fact that neither the air nor the vacuum are experienced, Hume rejects the idea of the latter, but not that of the former, on the grounds that "it has already been remark'd, that impressions can give rise to no ideas, but such as resemble them." This strongly implies that there is some sense in which the idea of air, even though it is an idea of something "that is not an object of the senses," does resemble some impression, and this in turn implies that we can represent that which is not experienced so long as such objects in some way resemble that which is experienced. That is, we can represent unobservable theoretical entities via perceptible models.[37]

Before turning to the task of investigating such models further, there is a complication here concerning *why* the air is not an object of the senses that it will be instructive to address. The complication is due to the fact that one natural answer to this question is that our idea of air is the idea of particles that are too small to be perceived. To take a toy example, we model our idea of air on the idea of billiard balls, and think of particles of air as having the same physical properties, exhibiting the same behaviors, and being governed by the same laws, except that they move much more quickly and are much smaller. The problem with this account of our representation of particles of air is that Hume explicitly and repeatedly denies that we can represent anything as being (or that anything can be) smaller than our *minima sensibilia*. That thesis implies that atoms of air are not imperceptibly small, but that we can in fact form adequate ideas of them merely by observing them under the proper conditions. So, these particles are not unobserv*able*, but rather only unobserv*ed*.

> 'Tis not for want of rays of light striking on our eyes, that the minute parts of distant bodies convey not any sensible impression; but because they are remov'd beyond that distance, at which their impressions were reduc'd to a *minimum*, and were incapable of any farther diminution. A microscope or telescope, which renders them visible, produces not any new rays of light, but only spreads those, which always flow'd from

> them; and by that means both gives parts to impressions, which to the naked eye appear simple and uncompounded, and advances to a *minimum*, what was formerly imperceptible.
>
> T 1.2.1.4; SBN 27–28

While particles of air might not be visible to us simply looking around the room, were we to look at a sample of air through a suitable microscope, we would be able to form adequate simple ideas of them (presuming that these particles were themselves simple). It would seem, then, that we might not use a perceptible model to represent the air at all. It might be that we represent the air using the senses alone: not via what we are *currently* observing, but what we have, will, or could observe in the proper conditions. That is, while the earlier passage appeared to indicate that the air and vacuum were similarly "not objects of the senses," perhaps there is an important difference in the way that this is true of each: the air is not a *current* objects of the senses, whereas the vacuum is not even so much as a *possible* such object. The difference between a representation of a vacuum and a representation of air the might reduce to the difference between that which unobservable and that which is observable after all.

There is reason, however, to think that this is not the case. While it may well be that particles of air are not unobservable in principle, the representation of such particles does not *typically* take the relatively straightforward form described above of a simple idea representing some simple impression.[38] The more typical mode of representation of such particles, even if these turn out to be observable, deploys exactly those resources needed for a more robustly theoretical-explanatory representation. That is, while we may be *able* to represent a particle of air by perceiving it, what we typically do is something very different (as evidenced by the theoretical apparatus that was necessary to first formulate the particulate theory of gases). Hume himself notices just this in a passage immediately following the one above.

> Nothing can be more minute, than some ideas, which form in the fancy; and images, which appear to the senses; since these are ideas and images perfectly simple and indivisible. The only defect of our senses is, that they give us disproportion'd images of things, and represent as minute and uncompounded what is really great and compos'd of a vast number of parts. This mistake we are not sensible of; but taking the impressions of those minute parts, which appear to the senses to be equal or nearly equal to the objects, and finding by reason, that there are other objects vastly more minute, we too hastily conclude, that these are inferior to any idea of our imagination or impression of our senses.
>
> T 1.2.1.5; SBN 28

The error on which Hume focuses here is one that results from our concluding from the discovery that what the senses have represented as simple is in fact complex that there are objects that are smaller than any *minima sensibilia*. What is of interest for current purposes, though, is how he takes this error to be discovered. It is not that the *senses* come to represent what is simple as simple; we do not, at least at first, come straightforwardly to *see* simple particles of air. Rather, we discover this error via "finding by reason, that there are other objects vastly more minute." That is, it is *reason* that plays the crucial role here, at least at first, of representing unobserved particles of air: the senses represent something as simple, but reason discovers it to be—forms a more accurate representation of it as being—complex. We do not, at least typically, represent particles of air via the senses, but rather experience reveals to us that what the senses present cannot be taken at face value, and combines with what the senses have provided to formulate a more adequate explanation that phenomenon.

Garrett describes the role of reason in Hume as follows. "A belief is *produced by reason* if and only if it results from an operation of the inferential faculty."[39] What the case of the imperceptible air suggests is that we can accept Garrett's description so long as we understand "the inferential faculty" to include inference to the best explanation. This suggestion will be born out in the rest of the examples to be presented, and in the next chapter's explication of Hume's distinction between reason, memory, and imagination. So consider next Hume's gloss on our experience of the world as consisting of three, rather than two, dimensions.

> 'Tis commonly allow'd by philosophers, that all bodies, which discover themselves to the eye, appear as if painted on a plane surface, and that their different degrees of remoteness from ourselves are discover'd more by reason than by the senses
>
> T 1.2.5.8; SBN 56

Here Hume notes in passing that it is not the image produced by senses alone that represents a three-dimensional space as three-dimensional, but that *reason* has an ineliminable contribution to make to that representation as well. What reason does in this case is the same as what it does in the case of the representation of unobservable air: it both discovers that what the senses provide requires further explanation, and contributes to forming a perceptible model to create a representation of an entirely new kind of object.[40] What Hume seems to have in mind in the case of the representation of a three-dimensional space is using a representation of a two-dimensional plane as the model to posit another such plane orthogonal to the first that would explain the observed behavior of certain colored points. (For example, when such-and-such a colored mass moves "farther away," we expect them to grow smaller and smaller until they finally disappear altogether, or that when one colored point "moves behind another," we see the first approach the second from the side, then disappear, and we then expect it to reappear on its opposite side, etc.).

Another example: in arguing against the infinite divisibility of the mind, Hume considers the idea of something a thousandth the size of a grain of sand, and while he does not explicitly refer to the role of reason in forming that representation, he does again note that the senses alone are inadequate to the task, and it is easy enough to trace the remainder of the work being done to reason using a few other texts.

> When you tell me of the thousandth and ten thousandth part of a grain of sand, I have a distinct idea of these numbers and of their different proportions; but the images, which I form in my mind to represent the things themselves, are nothing different from each other, nor inferior to that image, by which I represent the grain of sand itself, which is suppos'd so vastly to exceed them. [. . .] But whatever we may imagine of the thing, the idea of a grain of sand is not distinguishable, nor separable into twenty, much less into a thousand, ten thousand, or an infinite number of different ideas.
>
> T 1.2.1.3; SBN 27

Notice that Hume grants that we can form an idea of something a thousandth the size of a grain of sand—"I have a distinct idea of these numbers and of their different proportions"—but also holds that it is not the senses alone that produce that idea—"the images, which I form in my mind to represent the things themselves, are nothing different from each other." Of course, the distinct idea that we form of these numbers and their different proportions is that formed by considering the philosophical relation of "proportions in quantity or number" (T 1.1.5.6; SBN 14–15), and "except in very short numbers, or very limited portions of extension" (T 1.3.1.3; SBN 70) these depend on "a chain of reasoning" (T 1.3.1.5; SBN 71). So, here too we find reason (in this case, demonstrable rather than probable) playing a central role in moving beyond what is provided by the senses.

As mentioned earlier, though, while reason has an essential role to play in the construction of such models—it is what allows us to move *beyond* experience—this role is only made possible through the combination of reason with sensation. Whereas it is reason that discovers the need for a theoretical-explanatory posit and the specific form that such a posit will take, it is sensation that provides the model itself, and thus makes that which is posited intelligible. Specifically, as we saw in the passage concerning the imperceptible air, Hume repeatedly stresses that any such model must *resemble*, at least to some degree, that on which it is modeled. Here is another example of this sort of theoretical-explanatory activity in which Hume emphasizes both of these aspects: the role of reason in moving beyond the images provided by sensation, but also the role of sensation as the basis for reason's extrapolation.

> 'Tis universally allow'd by the writers on optics, that the eye at all times sees an equal number of physical points, and that a man on the top of a

> mountain has no larger an image presented to his senses, than when he is coop'd up in the narrowest court or chamber. 'Tis only by experience that he infers the greatness of the object from some peculiar qualities of the image; and this inference of the judgment he confounds with sensation, as is common on other occasions.
>
> T 1.3.9.11; SBN 112

Once again, the image that is provided by the senses is by itself inadequate to representational and explanatory task at hand: representing the vastness of the ocean. Thus, one "*infers* the greatness of the object from some peculiar qualities of the image." Of course, when Hume writes here that this inference is "only by experience" he does not mean that *only* experience is involved in drawing that inference, but as he does throughout the *Treatise*, he is here referencing the fact that all of our *probable reasoning*, our inferences from causes to effects and vice versa, are grounded in experience.[41] That is, that the inference that is drawn, the theoretical-explanatory entity that is posited, derives both its representational content and its evidential support from experience (here, for example, the experience of sailing the ocean). Hume continues by noting that, "a man has a more vivid conception of the vast extent of the ocean from the image he receives by the eye, [. . .] than merely from hearing the roaring of the waters," and then explains this fact by appeal to, "the resemblance betwixt the image and the object we infer."

> But as the inference is equally certain and immediate in both cases, this superior vivacity of our conception in one case can proceed from nothing but this, that in drawing an inference from the sight, beside the customary conjunction, there is also a resemblance betwixt the image and the object we infer, which strengthens the relation, and conveys the vivacity of the impressions to the related idea with an easier and more natural movement.
>
> T 1.3.9.11; SBN 112

Hume's point here is that the greater the resemblance between the model and the explanatory posit, the more easily and naturally we will accept the existence of the latter. It is plausible to suppose, though, that should this resemblance be reduced to nothing, not only would we no longer believe in the explanatory posit, but we would not even be able to so much as form an idea of it. In fact, we find Hume making this exact claim a few sections later.

> Without some degree of resemblance, as well as union, 'tis impossible there can be any reasoning: But as this resemblance admits of many degrees, the reasoning becomes proportionally more or less firm and certain. An experiment loses its force, when transfer'd to instances, which are not exactly resembling; tho' 'tis evident it may still retain as much as may be the foundation of probability, as long as there is any resemblance remaining.
>
> T 1.3.12.25; SBN 142

Here, as in the previous passage, we find Hume noting that greater degrees of resemblance correspond to greater firmness and certainty of belief, but he now also completes the picture of reasoning that we had supposed he was employing by noting that, "without some degree of resemblance [. . .] 'tis impossible that there can be any reasoning." This part of Hume's conception of the combination of reason and sensation in theoretical-explanatory activity is of crucial important because it is *precisely* the condition that he so often takes his predecessors to violate. That is, Hume finds such philosophers pretending to make some explanatory hypothesis in which there is *no* resemblance between the model and that which is posited. Consider the serious of questions that Hume puts to the philosopher that posits an immaterial soul.

> For how can an impression represent a substance, otherwise than by resembling it? And how can an impression resemble a substance, since, according to this philosophy, it is not a substance, and has *none* of the peculiar qualities or characteristics of a substance?
>
> T 1.4.5.4; SBN 233, emphasis mine

Notice that the problem with the idea of an immaterial soul is that it shares *none* of the peculiar qualities or characteristics of a substance, which suggests that if it shared *some* of these qualities, it could be used to do the explanatory work that is its intended purpose. Next, here is Hume writing about probable reasoning in general.

> Tho' the mind in its reasonings from causes or effects carries its view beyond those objects, which it sees or remembers, it must never lose sight of them entirely, nor reason merely upon its own ideas, without some mixture of impressions, or at least of ideas of the memory, which are equivalent to impressions.
>
> T 1.3.4.1; SBN 82

Notice that what Hume requires for a piece of causal reasoning to be legitimate is not that it not move beyond experience *at all*, but only that there be "some *mixture* of impressions" with its own ideas. That is, causal explanatory posits are legitimate just in case there is *some* resemblance between the model and that which is posited. Hume revisits this mixture of impressions with causal reasoning two sections later.

> Were there no mixture of any impression in our probable reasonings, the conclusion wou'd be entirely chimerical
>
> T 1.3.6.6; SBN 89

As we saw in examining Hume's explicit statements about the scope of theoretical activity in the previous section, a conclusion is not chimerical when it makes an explanatory posit that goes beyond experience at all, but only when it intemperately continues to seek explanation past the point of at which no further explanation is possible. Here we see what that point is: when "there is no mixture of any impression" in it. The complement to this thesis is that without reasoning, we could not move beyond experience at all.

> And were there no mixture of ideas, the action of the mind, in observing the relation, wou'd, properly speaking, be sensation, not reasoning.
> T 1.3.6.6; SBN 89

Finally, Hume combines these two theses to give an explicit endorsement of theoretical-explanatory posits:

> 'Tis necessary, that in all probable reasonings there be something present to the mind, either seen or remember'd; and that from this we infer something connected with it, which is not seen nor remember'd.
> T 1.3.6.6; SBN 89

It might be that all that Hume is after here is perfectly ordinary inductive inference: something present to the mind has in the past been constantly conjoined with something else, and so from this present impression, we infer the existence of a new instance of this other phenomenon, which is not seen nor remembered, but expected.[42] But there is nothing here that implies that we *must* read Hume that way. For all that Hume says in these passages, that which is neither seen nor remembered *could* be something new, a theoretical posit represented by recombining the material of the senses with a view towards the explanatory functions of reason. All that Hume *requires* in the passages above is that our probable reasonings have some mixture of impressions and reason. Inductive inference is allowable under that rule, but so would be the kind of robust theoretical-explanatory posit that I have been urging makes for a more charitable and richer understanding of Hume's scientific methodology.

In fact, while it is rare to find Hume explicitly addressing the notion of representing theoretical entities via a perceptible model, he does mention modeling here and there, almost always either to point out the need for a posit to resemble and differ from its model in some determinate ways or to deploy such a model by specifying such resemblances and differences himself. What we never find Hume doing is rejecting the notion of representation via modeling *per se*. Here, briefly, are a few examples of Hume's use of the term "model". In his arguments concerning the nature and origin of the idea of space Hume notes that,

> *Our* internal impressions are our passions, emotions, desires and aversions; none of which, I believe, will ever be asserted to be the model, from which the idea of space is deriv'd. There remains therefore nothing but the senses, which can convey to us this original impression.
> T 1.2.3.3; SBN 33–4

Notice that Hume accepts entirely the notion of forming a representation of space by modeling it on some impression, and rejects only the suggestion that it is our "internal impressions," impressions of reflection, that serve in this role (presumably because they do not resemble that idea in the proper way). Rather, he turns to the senses to provide "this original impression," i.e., the impression on which our idea of space is modeled.

This same dialectic appears reversed in Hume's consideration of the idea of causation wherein he rejects the suggestion that this idea is modeled on some impression of the power or influence that objects exert on each other, and instead turns to an impression of our own perceptions for his model.

> *Tho'* the several resembling instances, which give rise to the idea of power, have no influence on each other, and can never produce any new quality *in the object*, which can be the model of that idea, yet the *observation* of this resemblance produces a new impression *in the mind*, which is its real model.
>
> T 1.3.14.20; SBN 164–5

Again Hume clearly endorses the practice of modeling an idea with novel representational content on the matter given in perception, and rejects only the specific model proposed. While we will return to the idea of necessary connection in Chapter 5, and so will not delve into the details of this passage just now, it is worth noting that what is modeled on the new impression in the mind here is not just "more of the same," but rather this impression serves as the model for the idea of *power*, which differs from that impression at least insofar as it is applied at level below that of the determination of the mind that serves as its model. That is, whereas the model-impression is an impression *of the behavior of our ideas*, the idea of power that is modeled on it is an idea of a property belonging to the objects of those ideas. Thus, the use of a perceptible model here does exactly what I have been suggesting: it is used to generate novel representational content via the specification of determinate resemblances and differences between it and the target theoretical object.

During his discussion of our idea of body, Hume considers the questions whether our impressions resemble any quality found in external objects. Putting aside the details of his negative answer to that question, to which we will return in Chapter 4, it is revelatory to notice that he frames his conclusion in terms of modeling.

> 'Tis certain, that when different impressions of the same sense arise from any object, every one of these impressions has not a resembling quality existent in the object. For as the same object cannot, at the same time, be endow'd with different qualities of the same sense, and as the same quality cannot resemble impressions entirely different; it evidently follows, that many of our impressions have no external model or archetype.
>
> T 1.4.4.4; SBN 227

Because the qualities of objects cannot "resemble impressions entirely different," our impressions cannot be modeled on such objects. That inference at least weakly implies that the way that a model functions is by specifying determinate ways in which the posited entity both resembles and differs from its (perceptible) model. A posited entity that is specified only as being "entirely different" from some impression is modeled on that impression in name alone.

Hume repeats this critique of *faux* models farther along in the same section.

> For let us put two cases, *viz.* that of a man, who presses a stone, or any solid body, with his hand, and that of two stones, which press each other; 'twill readily be allow'd, that these two cases are not in every respect alike, but that in the former there is conjoin'd with the solidity, a feeling or sensation, of which there is no appearance in the latter. In order, therefore, to make these two cases alike, 'tis necessary to remove some part of the impression, which the man feels by his hand, or organ of sensation; and that being impossible in a simple impression, obliges us to remove the whole, and proves that this whole impression has no archetype or model in external objects.
>
> T 1.4.4.14; SBN 203–1

What Hume finds in considering these two cases is that there is nothing in external objects that at all resembles the feeling or sensation that accompanies the man's pushing his hand against a stone. He concludes from this that "this whole impression has no archetype or model in external objects." That is, modeling requires that the model and the novel representation resemble and differ from each other in determinate ways. Once again, Hume does not reject the notion of producing novel representational content via modeling, but only the specific way that such a model is purportedly deployed in this specific case. What these examples suggest, along with those presented earlier of Hume's clear but implicit use of perceptible models, is that Hume sees inference to the best explanation and the practice of forming representations of theoretical-explanatory entities via perceptible models (mixtures of impressions and reason) and a necessary component of the methodology of the science of human nature.[43]

Here is one final example from 1.3.16, "Of the reason of animals," where Hume appears to reason in just this way himself. In that section, Hume defends the thesis that animals have the same capacities for reason that humans do, and is explicit about his argument for this conclusion.

> 'Tis from the resemblance of the external actions of animals to those we ourselves perform, that we judge their internal likewise to resemble ours; and the same principle of reasoning, carry'd one step farther, will make us conclude that since our internal actions resemble each other, the causes, from which they are deriv'd, must also be resembling.
>
> T 1.3.16.3; SBN 176–7

It is important to note Hume's two-part procedure here. In the first part, Hume cites the kind of propaedeutic "more-of-the-same" reasoning that we mentioned above. Animals exhibit the same external behaviors that we do, we notice that in our own case this behavior is constantly conjoined with certain internal machinations, so we conclude that the internal lives of animals are similar ours. So far, that is mere inductive generalization. In the next step, however, we conclude from those inductively-established similarities of internal actions that "the *causes*, from which they are deriv'd, must also be resembling." Whereas our external and internal actions are observable, and so easily enough discovered or inferred to be similar, the *causes* of those actions are not. We do not simply introspect and find that the causes of our perceptions are this or that faculty. Such a conclusion must be earned by reasoning from the observed regularities of our perceptions to an inference to their best explanation. In this case that explanation comes in the form of positing a faculty of reasoning itself, which once established as the explanation for our own internal actions can likewise be posited as accounting for the inductively-established internal actions of animals. Thus, Hume appears here to countenance a kind of reasoning that does more than just infer that certain as of yet unobserved phenomena will resemble hitherto observed events: he also employs a kind of reasoning that explains these very regularities by appealing to their underlying cause.

In fact, the distinction between these two kinds of reasoning might be operative two paragraphs later that section as well.

> Here we must make a distinction betwixt those actions of animals, which are of a vulgar nature, and seem to be on a level with their common capacities, and those more extraordinary instances of sagacity, which they sometimes discover for their own preservation, and the propagation of their species. A dog, that avoids fire and precipices, that shuns strangers, and caresses his master, affords us an instance of the first kind. A bird, that chooses with such care and nicety the place and materials of her nest, and sits upon her eggs for a due time, and in a suitable season, with all the precaution that a chymist is capable of in the most delicate projection, furnishes us with a lively instance of the second.
>
> T 1.3.16.5; SBN 177

Dogs associate past encounters of fire with being burned and so avoid fire in the future. Hume classifies that instance of mere inductive reasoning as, "of a vulgar nature," "on the level with their common capacities." He contrasts that with "those more extraordinary instance of sagacity" that birds exhibit in making their nests, sitting on their eggs, etc. If that is where Hume stopped, we might just suppose that the difference between the dog's reasoning and the bird's is one of degree rather than kind (although it would then be strange why Hume would open this passage by writing that "we *must make a distinction*" between these two kinds of actions). Hume goes on, however, to liken the bird's reasoning to that of a *chemist*, who at the time of his writing would be the paradigm example of a theoretical scientist: a scientist who appeals to the underlying nature of the observed phenomena to explain their observed regularities. One way of understanding the difference that Hume here describes between the reasoning of a dog and that of a bird could well be that dogs employ mere induction whereas birds' reasoning is more sagacious precisely insofar as it demonstrates that they have a greater understanding of the mechanisms underlying their "own preservation and the propagation of their species."[44]

Even having shown that Hume endorses inference to the best explanation as a valid form of reasoning, however, still leaves open the question of what precisely is posited when theoretical-explanatory activity takes this form. As we saw in the previous section, what the science of human nature discovers is the powers, essences, and nature of the mind. Theoretical explanations function by suggesting that the observed particulars obey the empirical generalizations that they have been observed to obey *because* of the powers or essences of their underlying nature. That is, what such explanatory posits yield is a representation of the nature of the *substance* that underlies mental phenomena. That is (yet another) fairly radical proposal in the context of discussing the *Treatise* because, of course, Hume is supposed to be a notorious opponent of the very idea of substance. Really, though, Hume is only an opponent of a certain *conception* of substance: as an unknown *something* in which certain qualities are supposed to inhere (T 1.1.6.2; SBN 16), or as "any thing specifically different from ideas and impressions" (T 1.2.6.8–9; SBN 67–8), or as that which "has none of the peculiar qualities or characteristics" (T 1.4.5.4; SBN 233) of an impression. As we have seen, however, what is posited by the science of human nature is not a *mere* something, or anything specifically different from and entirely unlike perceptions. Rather, because the theoretical-explanatory posits of the science of human nature are represented via perceptible models that are a produced through a mixture of impressions and reason, they are necessarily determinate contentful representations of substances that resemble some impressions. Thus, none of Hume's objections to these bad-making features of his *predecessors'* conception of substance apply to those posits, and Hume is free to suppose that what the science of human nature discovers is the properly conceived substance of the mind. It is to an explication of that proper conception that I will turn in the next chapter.

Notes

1. Salmon, "Four Decades of Scientific Explanation," give an excellent overview of some of the most prominent refutations.
2. De Pierris, *Ideas, Evidence, and Method*, 150.
3. De Pierris, *Ideas, Evidence, and Method*, 153–4.
4. In fairness, Boyle's Law does also allow us to expand the scope of antecedently familiar laws of motion to include the behavior of gases, so the same example can be given an inductivist spin as well.

Nb. Boyle's Law, which states roughly that the pressure and volume of a gas stand in an inversely proportional relationship when the temperature of the gas is held constant, is an experimental approximation of the Ideal Gas Law, and so is more obviously a mere empirical generalization than the latter. Of course, it is Daniel Bernoulli in *Hydrodynamica*, published a year before the *Treatise*, who derives Boyle's Law by applying Newton's laws of motion to posited collections of molecules, thereby explaining the empirical regularity via an appeal to a theoretical posit.

5. While Locke does model the causal relations among corpuscles on the "impact causation" of observable objects such as billiard balls, Hume nonetheless takes him to have failed to properly model such corpuscles themselves. That is, Hume takes Locke to posit corpuscles that consist of primary qualities alone, which Hume argues in 1.4.4. it is impossible to conceive.
6. De Pierris, *Ideas, Evidence, and Method*, 12–13.
7. The caveat limiting Hume's realism to successful science is important. Schliesser, "Hume's Attack on Newton's Philosophy," provides one way of bringing this out. Schliesser argues that Hume distinguishes between three epistemic categories: demonstrations, proofs, and probabilities. Demonstrations are limited to relations of ideas; proofs and probabilities concern matters of fact. A matter of fact that is proved thereby becomes subjectively certain, whereas one that merely has evidence in its favor, no matter how much evidence, is only probable and therefore not certain. Schliesser argues that,

> at least four kinds of 'matters of fact' are susceptible to 'proofs': (1) claims about objects immediately present to senses and memory; (2) common sense (causal) claims; (3) results of some experiments in natural philosophy, especially if immediately present to eyes; (4) causal claims in moral sciences (*e.g.*, economics and politics).
>
> Schliesser, "Hume's Attack on Newton's Philosophy," 17.

I would suggest that what these four types of matters of fact have in common is that there appears to be no immediate need to *explain* them. For example, Schliesser cites Hume's claim that if a friend were to throw himself out of the window, he could know with certainly that he will fall (EHU 8.1.20) as an example of (2). The reason that we can be subjectively certain of that fact, that it admits of proof, is that we do not, at least in the moment, require any explanation of it. It appears, again in the moment, to be self-evident. The same is true of the objects that appear to the senses and memory (with their high degree of force and vivacity), experiments whose results are "immediately present", and causal claims in the moral sciences (e.g., Schliesser's example of fairly-priced goods always finding a buyer).

If it is true, though, that it is the lack of a demand for explanation of a matter of fact that makes it "proved" or subjectively certain, then whereas apparent self-evidence would be *one* way of achieving proof, the explanatory *completeness* at which science aims would be another—recall that Hume proposes "a compleat system of the sciences" (T Intro.6; SBN xvi). Of course, in the end, Hume does not take even his own pursuit of the science of human nature to have been *entirely* successful (T 1.4.7.14; SBN 272–3), and so not even it will achieve the status of proof. Still, what scientific realism requires is not that any *actual* science be provable, but only that science *when successful* be such. If the goal of science is explanatory completeness, and a matter of fact is proved when it needs no explanation, then successful science achieves proof. (It is worth noting that Hume does appear to think that such a complete system of science will be forthcoming in due time (T 1.4.7.14; SBN 273)). I will discuss the aims scientific explanation in more detail in Chapter 5.

8. Hsueh Qu points out that it is possible that we might discover that some inductively-established generalization standing in need of explanation cannot in fact be explained, which in turn might support the conclusion that induction is not valid after all. That is entirely compatible with the point here that insofar as one takes such a generalization as one's *explicandum*, one thereby proceeds under the working assumption that induction is valid. One could give up that assumption, but to do so would thereby be to cease seeking explanation for such generalizations.
9. For a different set of critiques of De Pierris's account of Hume's relation to Newton, see Hazony and Schliesser, "Newton and Hume," especially n68.
10. I continue to be concerned first and foremost with Hume's understanding of the science of human nature. As Boehm, "Hume's Foundational Project in the *Treatise*," argues, Hume conceives of all of the sciences (including mathematics, physics, morals, politics, religion, etc.) as depending on the science of human nature, and as Hazony, "Newtonian Explanatory Reduction and Hume's System of the Sciences," argues, this dependence amounts to an explanatory reduction of the phenomena observed and explained by these sciences to explanations comprising only the entities and laws discovered by the science of human nature. Hazony and Schliesser, "Newton and Hume," argue that this reduction should not be interpreted as *eliminativist* in part on the grounds that Hume takes himself to be giving an account *of* certain phenomena, which would not be possible if he took these *explicanda* to be "nothing but" the behaviors of their *explanans*. While I agree that explanatory reductionism need not be eliminavist *per se*, in Chapter 3 I will argue that the specific form that Hume's scientific realism takes does commit him to manifest phenomena's being mere appearance nonetheless.

    In Chapter 4, I will argue further that Hume is an anti-realist about the physical sciences insofar as he holds that the theoretical posit of *matter* is illegitimate (because it violates the condition that a theoretical posit be a mixture of impressions and reason). Schliesser, "Hume's Attack on Newton's Philosophy," also argues that Hume rejects the conclusions of Newton's physical science, but on the grounds that, unlike the explanations of Hume's science of human nature, Newton's explanations move beyond the firm epistemic ground of "common life," what is "provable," to conjectures regarding what is merely "probable". Since the evidence from Chapter 1 is that Hume's own distinctions are as conjectural as Newton's, it follows that Hume's rejection of Newton cannot proceed in this way.
11. T Intro.8; SBN xii.
12. Or as we will see Hume put it, contain a mixture of reason and impressions.
13. De Pierris, "Hume and Locke on Scientific Methodology," 279.
14. De Pierris's use of 'induction' is limited to what we would now call "enumerative induction", whereas there is a broader-scope use of "induction" that includes any form of reasoning that moves from observed experience to invisible principles or causes. (By contrast deduction moves from a general principle or cause to its effect.) Conceived broadly, enumerative induction is a form of induction because it makes implicit use of the principle of the uniformity of nature as its invisible principle, but so is inference to the best explanation, and the positing of unobservable theoretical entities. That said, I will continue to follow De Pierris in reserving "induction" and "inductivist" for enumerative induction, and contrasting these other forms of reasoning from the particular to the general with it.
15. De Pierris, "Hume and Locke on Scientific Methodology," 286. That explanation must ultimately be answerable to experience is what Garrett calls Methodological Empiricism. Garrett correctly reads this as aimed at rationalist explanations that *directly conflict* with experience, not as a blanket prohibition on theoretical-explanatory activity. Garrett, *Cognition and Commitment*, 30–1.

16. De Pierris, "Hume and Locke on Scientific Methodology," 292.
17. De Pierris, "Causation," 506. De Pierris, *Ideas, Evidence, and Method*, Chapters 4 and 5 argue that Hume's radical skepticism is a second-order skepticism concerning only the epistemological status of the principle of the uniformity of nature, the ultimate ungroundedness of which nonetheless leaves intact the norms governing common life and science. Nonetheless, as the quotation above brings out, even if Hume's meta-level skepticism about the principle of the uniformity of nature does not conflict with the norms of science, there is another kind of radical skepticism waiting in the wings that does. Namely, since De Pierris's Hume holds that observation is the only means of justifying scientific theorizing, he is also led to a radical skepticism regarding theoretical-explanatory posits. That might seem to be a relatively benign form of skepticism, but its consequences reach farther than I believe Hume would allow, a claim for which I will present evidence in the following two sections.
18. For example, see T 1.4.7.14; SBN 272.
19. De Pierris, "Hume and Locke on Scientific Methodology," 285–90.
20. Of course, one might respond that Hume's call to explain such exceptionless generalizations amounts to nothing more than subsuming one such generalization under a second that is wider in scope. For example, if universal gravitation were the result of the differing densities of the ether surrounding massive bodies, then one could cast it as an instance of the laws of motion. This way of proceeding, however, runs dangerously close to becoming the Deductive-Nomological account of scientific explanation of the Logical Positivists, and there is good reason to reject on Hume's behalf any such interpretation (even if the Positivist's mistakenly took themselves to be inspired by Hume).
21. Cf. Schliesser, "Hume's Newtonianism and Anti-Newtonianism."
22. One final note on this passage: this passage occurs as part of a note in which Hume insists that terms such a "*vis inertiae*" or "gravity" refer to only observed phenomena. That fact itself might appear to support the interpretation that Hume endorses Newton's casting of the ether as "mere hypothesis" because it appears to indicate that Hume holds that the content of theoretical terms is exhausted by a description of manifest phenomena. The context of this note is important, however. It comes at the conclusion of Hume's discussion of those philosophers who defend a "theory of the universal energy and operation of the Supreme Being." Hume's objections to that theory are that (a) the posited deity is unlike anything found in experience and so is incomprehensible, and (b) since we attribute precisely those powers to the posited deity that we already observe in manifest phenomena, these powers can serve no explanatory purpose. The point of the note, then, is to present two corresponding defenses of empirical science. The first emphasizes the close tie between science and experience; this is the point about the descriptive content of "*vis intertiae*" and "gravity". That these terms refer only to observed phenomena, though, does not imply that such phenomena do not themselves stand in need of explanation. Thus, the second point presents the form of a genuinely explanatory theory: Newton himself appeals to the ether to explain gravity. (Finally, Hume cautions that in both the divine and the scientific case acolytes can be too quick to accept mere hypotheses as established fact.)
23. My thanks to Jonathan Cottrell for calling my attention to the relevance of this passage here.
24. And let's be clear here: Hume is *very* worked up in this section. A moment later he releases the pendulum and lets it hit his opponent right between the eyes: "A man must be very sagacious, who could discover by reasoning, that crystal is the effect of heat, and ice of cold, without being previously acquainted with the operation of these qualities" (EHU 4.13; SBN 31–2).

25. For good measure, here is another passage wherein Hume appears at first sight to equate explanatory bedrock with mere empirical generalizations.

    > Wit, and a certain easy and disengag'd behaviour, are qualities *immediately agreeable* to others, and command their love and esteem. Some of these qualities produce satisfaction in others by particular *original* principles of human nature, which cannot be accounted for: Others may be resolv'd into principles, which are more general.
    >
    > THN 3.3.1.27; SBN 589

    Some qualities produce satisfaction because of some original principle of human nature. So far so good: the inductively-established regular co-occurrence of those qualities and satisfaction is explained via an appeal to the nature of the substance at hand, in this case us. However, it looks like Hume goes on to draw a contrast with qualities that stand in need of further explanation, and that he casts this explanation as their being subsumed under principles, "which are more general." If that is what Hume meant, that would certainly be strong evidence in favor of the inductivist position over scientific realism. Happily, it is clear from the context of this remark that this is not what Hume means at all! Hume begins the paragraph in which this passage appears by drawing a distinction between sentiments that arise, "from the mere species or appearance of characters and passions," and those that arise, "from reflections on their tendency to the happiness of mankind, and of particular persons." This distinction corresponds to the subsequent contrast between qualities that can be explained directly by appeal to human nature, and those must be resolved in more general principles. The generality of these principles does not consist in *their* being wider in scope, but instead in the fact that rather than being explained by an *immediate* relation of a person to a quality, they will be explained by an appeal to certain reflective principles, which principles themselves operate via the representation of general kinds. That is, we can only understand the goodness of some sentiments by reflecting on their general conduciveness to happiness to mankind. Rather than explain them by appealing to the nature of a particular quality and a particular person, we have to appeal to the relation of qualities of a certain kind to the species as a whole. In either case, what Hume is after is a substantial explanation (more on that in Chapter 3), not merely an empirical generalization of maximum scope. My thanks to Hsueh Qu for drawing my attention to the relevance of this passage here.
26. Qu, "Prescription, Description, and Hume's Experimental Method," is an interesting, but more equivocal investigation of how to interpret Hume's understanding of his scientific methodology. On the one hand, Qu appears to endorse an inductivist reading. For example, he understands Hume's claim from the *Enquiry* that philosophy discovers, "the secret springs and principles, but which the human mind is actuated in its operations," a passage that scientific realist interpreters such as the New Humeans seize upon with gusto, as describing nothing more than a "process of deriving general principles" (Qu, "Prescription, Description, and Hume's Experimental Method," 283). As Qu's investigation proceeds, however, things become less clear. As Qu reads it, Hume's methodology proceeds in *three* stages.

    > First is the taxonomic process of sorting and classifying a variety of observations; second, the derivation of general principles from these observations; third, the selection of suitable general principles on the basis of their theoretical merits.
    >
    > Qu, "Prescription, Description, and Hume's Experimental Method," 282.

Hopefully, it is clear that nothing about the scientific realism that I attribute to Hume is incompatible with the first two of these stages. The second describes the formulation of the inductively-established universal generalizations that are the *explicanda* of theoretical activity, and I will argue in the next chapter that the first is, as Qu recognizes, the necessary prelude to these. (That is, such generalizations by subsuming particular phenomena under general terms, a form of taxonomy.) Thus, a great deal hangs on what Qu means by "the selection of suitable general principles on the basis of their theoretical commitments". Here is what he means.

> I argue that Hume endorses three theoretical merits in his philosophy: empirical confirmation, explanatory power, and simplicity. Empirical confirmation is the degree to which a theory is supported by the observed evidence, and explanatory power is how much the theory explains. A theory is simpler than another if it postulates fewer things (ontological simplicity), or if its basic principles are fewer and/or more concise (syntactical simplicity). The idea is that theories which exhibit the best balance of these three theoretical merits are the ones that we should accept.
>
> Qu, "Prescription, Description, and Hume's Experimental Method," 284–5.

Empirical confirmation (depending how one understands the sense in which a theory can be "supported by the observed evidence") and explanatory power (which might be more accurately dubbed explanatory *scope*) are relatively uncontroversial as virtues that would make the cut on either an inductivist or scientific realist interpretation of Hume. What about simplicity? Well, syntactic simplicity amounts, roughly, to De Pierris's progress through generalizations of ever-increasing scope. Ontological simplicity, on the other hand, amounts to explaining, "all effects from the simplest and fewest causes"

(T Intro.8; SBN xvii), and as I noted earlier, the Logical Positivists understood DN as the means to avoiding causal explanations. Unfortunately for current purposes, Qu's distal aims in that paper are elsewhere, and he does not say much more about ontological simplicity.

27. For example, Owen, *Hume's Reason*, 65–6.
28. Of course, this depends on what exactly Hume means by "essence," "nature," and "powers," including how his understanding of those terms relates to that of his predecessors. I return to that issue in the following chapter.
29. A common way of understanding Hume's use of "explain" has been to understand it colloquially, as a synonym for "make clear" or "illustrate", as opposed to the more rigorous understanding that I advocate, which is more in line with the use to which "explain" is put in the philosophy of science. In a sense, this entire manuscript will be a case for understanding Hume as endorsing the more robust sense, and a variety of evidence for this interpretation will be presented. For example, we have already seen and will shortly see more examples demonstrating that Hume does, in fact, offer theoretical-explanatory posits as explanations throughout Book I. Furthermore, recent scholarship including De Pierris's account, but also, e.g., the opposition view of Schliesser, "Hume's Newtonianism and Anti-Newtonianism," also supports the claim that Hume employs a robust conception of scientific explanation, however we understand it.
30. An example of the first transgression would be the practice of those mechanical philosophers who took the conclusions of their *a priori* arguments about the nature of substance to be strong enough to override any empirical evidence to the contrary. An example of the second would be the Peripatetic philosophers that Hume scolds in 1.4.5 for their attempts to, "feign something unknown and

invisible," as that which underlies perceptual flux, the problem being not that such philosophers attempt to explain the behavior of our perceptions, but rather the "unintelligible something" to which they appeal to do so.

31. Hume does write that our experience of the reality of our most general and most refined principles is, "the reason of the mere vulgar," and that it requires, "no study at first to have discovered." His point there, though, is simply that the reasoning of the scientist of human nature is not different in kind from the reasoning of the vulgar. The inductivist takes this to be because both seek only generalizations of increasing scope. In Chapter 4 I will argue that, on the contrary, it is because the vulgar themselves are already engaged in a kind of theoretical-explanatory activity. That argument will require a detailed examination of Hume's portrayal of the vulgar's understanding of body.
32. The relation of the science of human nature, or "true philosophy," to the systems of the vulgar and the false philosophy will be the subject of the Chapters 3 and 4.
33. To be clear, merely *forming* a perceptible model ("a mixture of reason and impressions") or *drawing* some analogy is not sufficient for taking that which is thereby represented to exist. Belief in the existence of that which is pictured in this way can only be justified by the *explanatory force* of such a theoretical posit. So, merely picturing an atom as having a structure like the solar system is not what warrants believing in atoms, but rather that doing so *explains*, e.g., the Rydberg formula for the spectral emission lines of atomic hydrogen. My thanks to Miren Boehm for prompting this clarification.
34. Hume describes analogy as a species of probable, i.e., causal, reasoning at T 1.3.12.25; SBN 142, and there emphasizes its dependence on resemblance, i.e., on the possibility of employing perceptible models in theoretical explanation. More on analogy in Chapter 5.
35. Although Hume appears to become more optimistic on this score later in life (EHU 1.8–9; SBN 10–11). Notice in these later passages once again it is *accuracy* that is the end product of science, which just so happens, as Hume then sees it, to serve the practical needs of the arts.
36. I postpone a full explication of Hume's distinction between reason and imagination until the next chapter.
37. Where an object is supposed to differ *entirely* from any experience, though, we can form no idea of it. It is the latter thesis that Hume uses here to argue that we have no idea of vacuum, but as we will see in later chapters, it also plays a crucial role in his arguments against the ideas of external existence, necessary connection, and personal identity. See Chapters 4 and 6.
38. Of course, our investigation of the nature of simple ideas in the previous chapter goes a long way towards eliminating this interpretation as a live possibility.
39. Garrett, "A Small Tincture of Pyrrhomism," 80.
40. Kail also notices the striking role that Hume assigns to reason in this passage, and argues that because it can be neither demonstrable reasoning, nor probable reasoning (enumerative induction) that does this work, Hume must not mean what he says here.

> The passage just quoted comes from Hume's discussion of space and time, and although he hints that we discover distance through reason he does not explicitly address the topic of how the three-dimensional trick is supposed to be done until 'Of skepticism with regard to the senses'. By that stage, however, it is difficult to see how it can be 'discover'd by reason' at all." [. . .] The problem is that one cannot see, given Hume's conception of demonstrative and probable reason, how it would be possible to reason from continuity to distinctness [. . .] Since we do not detect 'outness' by the senses or reason our experience of it is a projection.
>
> Kail, *Projection and Realism*, 54–5.

Kail's argument would only be valid if demonstrable reasoning and probable reasoning (as Kail understands it) were the only possibilities available for understanding reason's role here. As I have been arguing, there is another option available—inference to the best explanation—that can do this work. Of course, Kail is right that by the time Hume gets to 1.4.2, Hume's claim here becomes more problematic, and in fact, I agree that Hume ultimately rejects it. His *grounds* for doing so, however, are not the ones that Kail cites. In fact, in both of these first two examples—of the imperceptible air and the third spatial dimension—as well as in some of the following ones—e.g., of the vastness of the ocean—Hume will ultimately reject the theoretical posit at hand. As it turns out, all three of these cases implicitly depend on representing the continued and distinct existence of some of our perceptions, which, as we will see in Chapter 4, Hume argues is impossible. As we will also see there, what happens in each of these cases is that the biases of the imagination lead us to *take* there to be a certain inductively-established universal regularity of experience—first the constancy, and then the coherence of our perceptions—when there is in fact none. Thus, reason operates to attempt to explain this regularity via deploying a perceptible model to form a theoretical posit, but is eventually thwarted because of its *explicanda* are merely illusory. These examples nonetheless stand as paradigms of the proper *methodology* of theoretical explanations, precisely because what ultimately dooms them to failure is *not* the role that reason play in them, but that they illicitly rely on the deliverances of the imagination.

41. See, for example, T 1.3.4.1; SBN 82–3, 1.3.6.2; SBN 87, T 1.3.6.6; SBN 89, T 1.3.6.8; SBN 90, T 1.3.7.6; SBN 97, 1.3.8.8; SBN 101–2.
42. This form of reasoning, which Hume calls argument from analogy, will be the subject of part of Chapter 5. I there present evidence that Hume does not take this form of argument to have any explanatory value of its own. Explanations that appeal to the "nature" of the substance underlying experience, by contrast, do have such force. As such substantial explanations themselves rely on the use of a perceptible model to form a theoretical representations formed via an "analogical" extension of experience, the vocabulary here can easily and unfortunately become confused.
43. A few more examples from texts other than the *Treatise*. Hume laments that British orators did not more closely resemble their ancient model (El 17). Hume suggests modeling one's behavior on that of a character that one approves and becoming "well acquainted with those particulars, in which his own character deviates from this model" (Sc 31). Hume reports on the religious practitioners of "idolatrous nations" that they reduce, "heavenly objects to the model of thing below, they may represent one god as the prince or supreme magistrate of the rest, who, though of the same nature, rules them with an authority, like that which an early sovereign exercises over his subjects and vassals" (N 6.5). In the *Dialogues* Hume repeatedly questions the validity of representing the Deity by taking our own minds as a perceptible model, e.g., at D 2.19, D 3.12, and D 7.17.
44. Hume spends the remainder of this section discussing only the dog's reasoning, and writes of it that,

> Beasts certainly never perceive any real connexion among objects. 'Tis therefore by experience they infer one from another. They can never by any arguments form a general conclusion, that those objects, of which they have had no experience, resemble those of which they have. 'Tis therefore by means of custom alone, that experience operates upon them.
>
> T 1.3.16.8; SBN 178

The reasoning of dogs is the result of mere custom, and because it does not represent any items as resembling one another (via the use of general terms), it does not represent the real connections among objects. Hume never draws the implicit contrast with humans and birds because his purpose in this section is to emphasize the *similar* role that custom and habit play in the process of belief formation in animals and humans. It is nonetheless there for the drawing: humans and birds represent real connections among objects by representing them as resembling (in humans, this is through the use of general terms). We have already touched on this thesis in the previous chapter, and will return to it in detail in the next one.

3 The Course of Science

Substance, Language, and Reason

In the previous chapter, I defended an interpretation of Hume's understanding scientific explanation according to which science aims at explaining the inductively-established universal regularities of perception via an appeal to the underlying substance of the mind. Such regularities, I argued, are discovered by reason, and the posits that are meant to explain these are a product of reason's using the deliverances of the senses as a model to form representations with novel content. Those theses have now left me with a number of expository debts that I intend to cash in this chapter.

The first of these is regarding my claim that the science of human nature discovers the nature, powers, and essence of the substance that underlies the human mind. Defending that claim requires an exposition of Hume's understanding of substance, which I undertake by way of a close reading of 1.1.6, "Of modes and substances," and other relevant texts. What I conclude from that reading is that Hume is not opposed to the notion of substance or substantial explanation *per se*, but only the use to which his predecessors put it, as a *via negativa* that can only be described as being entirely unlike anything in experience. As I stressed in the previous chapter, Hume requires theoretical representations to resemble and differ from some antecedently contentful model in determinate ways, and the notion of substance as a *via negativa* violates this important condition. By contrast, as Hume understands it, the proper representation of substance is formed by employing a general term with "a uniting [causal] principle as its foundation." That is, to represent substance is to represent certain properties as all stemming from a common cause by subsuming those properties under a general term, thereby representing their common cause as being of a certain determinate kind. For example, we represent the substance "gold" by representing its substantial properties (its color, malleability, solubility in *aqua regia*, etc.) as all stemming from its atomic properties. Or we represent impressions as being of a distinct kind by taking their high degree of force and vivacity to stem from their nature as mental originals.[1]

As I will show, the distinction between a general term that represents a substance (with a uniting principle as its foundation) and one that represents a mere mode, corresponds to the distinction articulated in Chapter 1 between the Ontological Interpretation and the Nominalist Interpretation of Hume's theory of general representation. Thus, the time will have come to cash the promissory note issued there of showing that while Hume is a Nominalist with respect to the general terms of the language of the vulgar, he is takes the terms of the scientist of human nature to reflect the nature, powers, and essence of the substance underlying the human mind. In fact, what I will argue is that it is by *replacing* the language of the vulgar with the language of science because of its greater explanatory force that we represent the general terms of the latter as more accurately reflecting the substance of the mind.

The greater explanatory force of the language of science is important because replacing one language with another can be done for a variety of reasons, not all of which are ontologically committing. In the final section of this chapter, I will argue that explanatory force does carry such a commitment for Hume because he takes it to reliably track truth. That thesis comes via an explication of Hume's understanding of reason, which I will argue includes an endorsement of inference to the best explanation. While one of the roles of reason is to discover the universal regularities in experience, another of its roles is to explain those regularities, which it does by recombining the deliverances of the senses to form theoretical posits with novel representational content. As such, reason is similar to the imagination, but importantly differs from imagination insofar as its representations essentially function to explain, are answerable to experience, and reliably track the truth. As I said, then, the debts of the previous two chapters have placed a number of items on the agenda for this one, and it is to those that I now turn, beginning with Hume's understanding of substance.

Substance

Hume devotes a small but important early section of Book I, 1.1.6, "Of modes and substances," to accounting for the difference between substances and modes, and my plan here is to work through that section in some detail. Hume begins there by challenging those philosophers who have relied on the specious distinction between substances and accidents to produce the impression from which the idea of substance is derived. After demonstrating that they will not be able to do this because *ex hypothesi* this notion of substance is one according to which a substance is that in which perceptible qualities inhere, but which itself is entirely different from such qualities, Hume concludes:

> We have, therefore, no idea of substance, distinct from that of a collection of particular qualities, nor have we any other meaning when we either talk or reason concerning it.
>
> T 1.1.6.1; SBN 16

Having demonstrated that our idea of substance is nothing over and above the idea of a collection of particular qualities, Hume recognizes that this leaves him with an explanatory burden insofar as this appears to undermine the intuitively plausible distinction between substances and modes.

> The ideas of substance as well as that of a mode, is nothing but a collection of simple ideas, that are united by the imagination, and have a particular name assign'd them.
>
> T 1.1.6.2 SBN 16[2]

The reference to having a particular name assign'd to them is an anticipation of the theory of general ideas that Hume will introduce in the following section. In a moment we will consider exactly how Hume puts that theory to service in this context. What is important now is just that at this point in the section, Hume appears to be claiming that the idea of a substance and the idea of a mode are both just instances of the theory of general representation, and so is in danger of losing the distinction between them. Thus, Hume's next move is to recover that distinction in a series of dense and obscure claims that require unpacking. He begins with the idea of substance.

> But the difference betwixt these ideas consists in this, that the particular qualities, which form a substance, are commonly referr'd to an unknown *something*, in which they are suppos'd to inhere; or granting this fiction shou'd not take place, are at least suppos'd to be closely and inseparably connected by the relations of contiguity and causation.
>
> T 1.1.6.2; SBN 16

The clause here about the idea of substance being referred to an unknown something is clearly a reference to the account of substance that Hume has just rejected, so we can excise it for the sake of clarity.

> the particular qualities, which form a substance [. . .] are at least suppos'd to be closely and inseparably connected by the relations of contiguity and causation.
>
> T 1.1.6.2; SBN 16

The idea of substance, properly conceived, is the idea of some particular qualities, assigned the same name, that are "suppos'd to be closely and inseparably connected by the relations of contiguity and causation." The relation of contiguity that Hume refers to here is the contiguous appearance of each of the qualities at hand, but that turns out to be a non-starter because as Hume notes a moment later, it is easy enough to imagine a scenario in which a mode displays this same relation.

> The simple ideas of which modes are form'd, either represent qualities, which are not united by contiguity and causation, but are dispers'd in

> different subjects; or if they be all united together, the uniting principle is not regarded as the foundation of the complex idea.
>
> T 1.1.6.3; SBN 17

As Hume notes here, while modes are typically "dispers'd in different subjects," they need not be, and when they are not, what distinguishes a mode from a substance is that with respect to the former, "the uniting principle is not regarded as the *foundation* of the complex idea." The "complex idea" here refers to the general representation that is used to represent the substance, and it is striking that Hume takes such ideas to operate via a uniting principle that serves as their foundation. That is surprising because as we saw in Chapter 1, in the section immediately following this one, Hume presents his theory of general representation, and there portrays general representation as depending on only resemblance relations, making no mention of the uniting principles serving as the foundation of such representations. Thus, we can conclude that this is not a feature of general representation *per se*, but that having a uniting principle serve as the foundation of a general representation is a feature unique to the representation of a substance. That accords with the dialectic that we have seen unfold in this section according to which both substances and modes are represented via general terms, but that there is a distinction between these two kinds of representation nonetheless. The representation of a substance, unlike that of a mode, has a uniting principle and a foundation. That is, it is more than a mere collection of qualities: it is a collection of qualities that together compose the essence, powers, or nature of some real thing.

That is getting a bit ahead of ourselves. Where things stand in our exegesis to this point is as follows. What we have just discovered is that even in cases in which a certain quality can be found among all of the ideas that compose the revival set of some general term, that quality will not be counted as belonging to the substance represented by that general term unless it meets some further criterion: falling under the uniting principle that is the foundation of the complex idea. Having eliminated contiguity as a reliable criterion by which to distinguish modes from substance, we are left with the causal relations that Hume mentions as the only remaining contender for playing the role of this uniting principle. What I now want to show is that the most plausible such causal principle is a theoretical-explanatory one.

At this point, we have seen Hume cast substances and modes as both being collections of certain qualities. Hume then distinguishes these by noting that the former, but not the latter, stand in certain relations of contiguity and causation. He then quickly recants the importance of the former relation, leaving causation as the foundation of our proper representation of substance. Without further specification, however, it is unclear precisely *which* causal relation Hume has in mind here. Here are three possibilities. The ideas that compose the idea of a substance might be supposed to:

(1) stand in causal relations to *each other* (Q_1 causes Q_2),
(2) collectively play a certain *causal role* (Q_1 and Q_2 together cause E).
(3) have a *common causal explanation* (the existence of both Q_1 and Q_2 are explained by the nature of substance S),

By way of explaining each of these three possibilities, consider Hume's own example of the qualities of the substance gold. After noticing that what qualities we take to be substantial ones can change as we discover additional qualities that stand in the appropriate causal relation (whatever it may be), Hume writes:

> Thus, our idea of gold may be at first be a yellow colour, weight, malleableness, fusibility; but upon the discovery of its dissolubility in *aqua regia*, we join that to the other qualities, and suppose it to belong to the substance as much as if its idea had from the beginning made a part of the compound one.
>
> T 1.1.6.2; SBN 16[3]

Our idea of gold is the idea of a certain kind of substance, and that idea is composed of the ideas of certain qualities: color, weight, malleability, fusibility, and eventually solubility in *aqua regia*. Every sample of gold, though, will also have certain other qualities—e.g., a shape—and so we can ask: what is it that makes the former qualities *substantial* qualities of gold, rather than modes? As we saw above, Hume's answer is that the substantial qualities, but not the modes, stand in a certain causal relation, and so the next question to be considered is precisely *which* casual relation this is. For example, to take (1) to be the relevant causal relation would be to suppose that it is in virtue of gold's color causing its weight or its malleability causing its fusibility, etc. that these qualities are substantial qualities. That does not seem very plausible at all. For one, Hume is clear that causal relations are never simultaneous, and any and all of these qualities might well be perceived simultaneously. For another, we simply do not think of these qualities as causing one another, despite our thinking of them together constituting the substance gold.

Taking (2) to be the relevant causal relation is to suppose that in thinking of certain qualities as substantial qualities, we think of those qualities as playing a *unified* causal role. For example, we might think that it is the color and malleability of gold that are together the cause of a particular brick of gold's falling the ground when dropped. Here again we encounter a conflict between the suggestion at hand and our common conception of the relation of these properties to one another. For example, we do not, in fact, think of the color of gold, much less its fusibility or its solubility in *aqua regia*, as playing any causal role in a brick of gold's falling to ground. As Hume explicitly notes in 1.3.13 "Of unphilosophical probability," just as we need to distinguish between substantial qualities and modal qualities, we must likewise distinguish those causes that are essential to particular event's occurrence and those that are merely accidental to it.

> In almost all kinds of causes there is a complication of circumstances, of which some are essential, and other superfluous; some are absolutely requisite to the production of the effect, and others are only conjoin'd by accident.
>
> T 1.3.13.9; SBN 148

The color of gold is "only conjoin'd by accident" to the brick of gold's falling when dropped, and so is a superfluous rather than an essential cause. Since its color is a substantial quality of gold, but is not always causally efficacious, then it cannot be that causal efficacy is a necessary condition of a quality's being a substantial quality. Furthermore, because each of these qualities will be causally efficacious in different circumstances, rather than being united by their causal roles, they seem rather to be divided by these. Malleability is a quality that is more closely related to other metals than it is to the yellow color of gold. So, it does not appear to be their collective causal role that unites these properties into a representation of substance, since they do not appear to have such a collective causal role at all.

That leaves (3), and I believe that we have good reason to think that Hume has just such a causal relation in mind as that which serves as the "uniting principle" and "foundation" of the general representation of substances. Taking (3) to be the relevant causal relation is to suppose that in thinking of certain qualities as substantial qualities, we suppose that the causal explanation of each quality will make reference to some single thing. It is easy enough, for example, (for us) to consider *all* of the qualities that Hume lists as substantial qualities of gold—its color, weight, malleability, fusibility, and solubility in *aqua regia*—as explained by its atomic structure. Here again is Hume's proposal for what makes a quality a substantial quality.

> the particular qualities, which form a substance [. . .] are at least suppos'd to be closely and inseparably connected by the relations of contiguity and causation.
>
> T 1.1.6.2; SBN 16

The current suggestion is that to suppose that gold's color, weight, malleability, etc., are "closely and inseparably connected by the relations of contiguity and causation" is to suppose that there is a single causal explanation tying these qualities together. In the specific case of gold, we have such an explanation ready at hand—it is gold's atomic structure that accounts for all of these properties—but similar explanations can be given, at least provisionally, for anything that we suppose to be a substance.

Of course, even to consider this option as a live possibility for Hume, this single thing cannot be a mere "unknown something," but will have to be (at least potentially) representable by a perceptible model. It cannot be, that is, that we take the substance gold to be a *via negativa* in which the relevant qualities inhere. Thus, in 1.4.3 "Of the antient philosophy," Hume castigates the Peripatetics for holding that substance is something "unknown and invisible."

> In order to reconcile which contradictions the imagination is apt to feign something unknown and invisible, which it supposes to continue the same under all these variations; and this unintelligible something it calls a *substance*, or *original and first matter.*
>
> T 1.4.3.4; SBN 220

What is objectionable here is not seeking an explanation for the constancy and coherence of our perceptions (although more on that in the next chapter), nor in making an appeal to substance as the means of that explanation, but rather in the appeal to an entirely unintelligible *understanding* of substance. Later in that same section, discussing how this empty notion of substance leads to an equally empty notion of qualities as accidents that inhere in such a substance, Hume contrasts the specious reasoning of the Peripatetics with the causal reasoning of the scientist of human nature.

> For having never discover'd any of these sensible qualities, where, for the reasons above-mention'd, we did not likewise fancy a substance to exist; the same habit, which makes us infer a connection betwixt cause and effect, makes us here infer a dependance of every quality on the unknown substance.
>
> T 1.4.3.7; SBN 222

Where the Peripatetic illegitimately infers that accidents inhere in an unknown substance, the scientist of human nature legitimately infers that the qualities of a substance (represented via a perceptible model) are related via a common *cause*. Again, it is not positing that a substance exists and can explain the behaviors of certain qualities that is specious, but only misunderstanding the nature of that substance and its relation to these qualities that is. Hume makes all of this explicit in the opening paragraph of the following section, 1.4.4 "Of the modern philosophy."

> I must distinguish in the imagination betwixt the principles which are permanent, irresistible, and universal; such as the customary transition from causes to effects, and from effects to causes: And the principles, which are changeable, weak, and irregular; such as those I have just now taken notice of. The former are the foundation of all of our thoughts and actions, so that upon their removal human nature must immediately perish and go to ruin. The latter are neither unavoidable

> to mankind, nor necessary, or so much as useful in the conduct of life [. . .] For this reason the former are receiv'd by philosophy and the latter rejected.
>
> T 1.4.4.1; SBN 225

Philosophically sound practice relies on the kinds of causal explanations that we have seen Hume take to provide the foundation for the proper notion of substance. Where the Peripatetics go wrong is in attempting to base their explanations on notions that are entirely empty. Hume makes this contrast even clearer in the continuation of this passage in which he draws an analogy between the reasoning of scientist of human nature and that of the Peripatetics to two people reasoning about what is in the darkness around them.

> One who concludes somebody to be near him, when he hears an articulate voice in the dark, reasons justly and naturally [. . .] But one, who is tormented he knows not why, with the apprehension of spectres in the dark, may, perhaps, be said to reason, and to reason naturally too: But then it must be in the same sense, that a malady is said to be natural; as arising from natural causes, tho' it be contrary to health, the most agreeable and most natural situation of men.
>
> T 1.4.4.1; SBN 225–6[4]

The proper remedy to the Peripatetics' appeal to substance as an unknown something—a spectre in the dark—as a specious explanation is not to abandon the natural and legitimate practice of substantial explanation *per se*, but rather is to ensure that this is done properly. It is perfectly legitimate to explain an articulate voice in the dark by supposing that voice to belong to *a person*. What is illegitimate is to seek explanations where none are needed and offer as an explanation a mere unknown something. Hume summarizes his critique of the Peripatetics as follows.

> But these philosophers carry their fictions still farther in their sentiments concerning *occult qualities*, and both suppose a substance supporting, which they do not understand, and an accident supported, of which they as imperfect an idea. The whole system, therefore, is entirely incomprehensible, and yet is deriv'd from principles as natural as any of those above-explain'd.
>
> T 1.4.3.8; SBN 222, emphasis mine

The problem with the peripatetic philosophy is its reliance on spurious explanations, but as the final clause here again makes clear, this is not a condemnation of explanation in general—the impulse here is a natural and perfectly legitimate one—but only of explanatory posits that are ultimately occult, not understood, imperfect, and incomprehensible.

The proscription on such specious explanations, however, does not prevent us from thinking of the substance gold as itself being the common causal explanation of each of these properties, so long as we are able properly to give content to this supposition. It is no accident, for example that in taking gold to be a substance with a certain atomic structure, we employ the Rutherford-Bohr model of the atom, which takes the solar system as its perceptible model, and specifies determinate ways in which an atom both resembles and differs from this. (Like the solar system, atoms consist of a massive central body being orbited by smaller bodies. The orbits of electrons are circular rather than elliptical, and the result of electrostatic rather than gravitational forces, etc.)

Thus, my suggestion is that in thinking of certain qualities, but not others, as being substantial qualities, we think of those qualities as being the result of some underlying essence or nature of the substance at hand. In fact, Hume proposes just such a criterion in presenting what it is that makes us take a particular object to be identical across time (which, as Hume notes, will itself depend on thinking of such an object as being of a certain kind of substance).

> Whenever we discover such a perfect resemblance [of the former and present qualities of the object], we consider, whether it be common in that species of object; whether possibly or probably any cause cou'd operate in producing the change and resemblance; and according as we determine concerning these causes and effects, we form our judgment concerning the identity of the object.
>
> T 1.3.2.2; SBN 74

We think of an object as being identical across time just in case we think of it is as being of a certain "species of object" in which "any cause cou'd operate in producing the change and resemblance." That is, we think of an object as being identical across time just in case we think that there is some univocal causal explanation for how objects of that kind come to have the qualities that they do. Supposing that representing an object as being of a certain species is equivalent to representing it as being a certain kind of substance, we can conclude that to represent something as a substance is to represent there being some univocal causal explanation for its having the qualities that it does.

To return to the contrast between substances and modes, a modal property is one that is not supposed to relate to the underlying nature of the substance at hand. Hume's own examples of modes are of a dance and beauty.

> The simple ideas of which modes are formed, either represent qualities, which are not united by contiguity and causation, but are dispers'd in different subjects; or if they be all united together, the uniting principle is not regarded as the foundation of the complex idea. The idea of a

> dance is an instance of the first kind of modes; that of beauty of the second. The reason is obvious, why such complex ideas cannot receive any new idea, without changing the name, which distinguishes the mode.
>
> T 1.1.6.3; SBN 17

A dance is not a substance because it is "dispers'd in different subjects." Of course, this presupposes that we have a working notion of the identity of different subjects, and that is precisely the work that an account of substances provides. Plausibly, though, the causal explanation of a dance would be given in terms of the conglomerate causal explanations of the individual dancers. That is, "a dance" is most plausibly construed as a nominalization of the verb "dancing," which applies most readily to persons, which in this context would be the subjects, or substances, at hand, the substantial qualities of which we take to be explained by some uniting principle.

By contrast, beauty is not a substance because while it is "united together, the uniting principle is not regarded as a foundation of the complex idea." That is, while it is a single object that is beautiful—again presupposing that we have a notion of the identity of the substance at hand—that object is not the object that it is in virtue of being beautiful. For example, a sculpture is considered a substance in virtue of the causal structure of that object, whereas its being beautiful is not attributable to this same structure (at least not *per se*). In accounting for the causal structure of the sculpture, we would make reference to the properties of the material of which it is composed: its color, hardness, malleability, rigidity, etc. We would expect these qualities to have a common causal explanation that would make reference to the nature of this substance—e.g., the properties of clay or stone—but we would not expect the beauty of the statue to require a similar explanation.

Finally, unlike a substance, a mode "cannot receive any new idea" because a quality that is a mode is not "closely and inseparably connected" to any other qualities. By contrast to a substance, the general term that is annexed to the qualities that comprise a mode do not have a uniting principle that provide their foundation. They are simply those qualities that are called by a certain name. Should we change what qualities these are, we change the meaning of term, and thus refer to a different mode. What this implies, though, is that the opposite is true of substances. The general term that represents a substance does not merely represent a list of qualities, but represents these qualities *as* being connected via some uniting principle or foundation. That is, these qualities *belong to* a certain substance, which is why we can discover that *further* qualities also belong to that substance, or that some qualities that we took to be substantial are not so.

We are now in a position to add a further thesis to our picture of Hume's understanding of the science of human nature. We observe that certain mental particulars obey certain empirical generalities. This fact demands an explanation. That explanation takes the form of a theoretical posit, the representation of which requires a combination of both experience and reason, or what I have been calling a perceptible model, which resembles and differs from that which is posited in determinate ways. Such a theoretical posit performs its explanatory role by casting the substance as having as an essence, power, or nature such that it obeys the empirical regularities that it does. All of which is to say that the explanations employed by the science of human nature are essentially ontological explanations. It is because a gas *is* a collection of atoms obeying the known laws of motion that it obeys the Ideal Gas Law. It is because gold *is* an atom with a certain atomic structure that is yellow, malleable, and dissolvable in *aqua regia*. It is because our complex perceptions *are composed of* simple ones that we can explain the variety of human thought given the paucity of experience, etc.[5]

While the focus of the bulk of this section has been on *what* we represent in representing substance—certain qualities as having a common causal explanation—as I stressed earlier, equally important to Hume's account is *how* we represent these. As we have seen, we represent such substances by employing a general term with uniting causal principle as its foundation. By contrast, we represent a mode when we employ a general term without such a principle. To say that much, however, leaves unsaid what it is for a general term to have a uniting causal principle as its foundation. It is to the business of filling that lacuna to which I turn in the next section.

It is worth signposting at this point that turning to the issue of how we represent substances is simultaneously a return to a promissory note that I have left uncashed since the end of Chapter 1. By the end of that chapter, I had concluded that representing both the simple-complex and impression-idea distinctions requires forming a representation the content of which necessarily moves beyond what is allowed by a thoroughly Nominalist Interpretation of Hume's theory of general representation (according to which what is represented by a general idea is merely a reflection of the associative tendencies of the human mind). What I argued there is that the explanatory power of these distinctions requires understanding the representations associated with their terms as carrying an ontological commitment to the real existence, qualities, and relations of that which they represent. For example, simple ideas are *really different* from complex ones, and are not just those ideas represented by *associating* certain complex ideas with one another. That account of these two distinctions, like the account of substance offered above, demands that there be some way to undertake an ontological commitment using representational resources not contained in Hume's theory of general representation *per se*. We must be able to represent simple and complex perceptions, impressions and ideas, and substances such as gold or lead as *really* different from one another, as having a uniting principle as their foundation, even in the face of no such mechanism being specified by Hume's theory of general representation. It is to detailing that mechanism, and thereby making an initial payment on these promissory notes, to which I now turn.

The Language of Science

Wilfrid Sellars once described the task of the scientific realist with respect to the language of physical science as follows.

> [T]he Scientific Realist need only argue [. . .] that in principle this language could *replace* the common-sense framework in *all* its roles, with the result that the idea that scientific theory enables a more adequate picturing of the world could be taken at its face value.[6]

What I want to suggest is that Hume's pursuit of the science of human nature reflects a similar brand of scientific realism: it is by privileging one language over another, because of its greater explanatory force, that we picture the world as containing certain natures, essences, and powers. In Hume's case, the common-sense framework takes the form of the language of the vulgar, and this language comes to be replaced by his own theoretical apparatus, including and especially the kinds of distinctions that were our focus in the first chapter.[7]

To begin, consider an example of a general term used by the vulgar, say, 'dog'. For that general term to represent a substance, the vulgar must take it to have a uniting principle as its foundation, i.e., they must take the properties of dogs to stem from a common cause. What the science of human nature discovers, however, is that what the vulgar take this common cause to be—certain perceptions that exist distinct from our experience of them and that continue to exist when we are not perceiving them—is not, in fact, their common cause at all, and in fact, there is no such common cause. The qualities that the vulgar take to be the result of the substance, dogs, do not in fact have a uniting causal principle as their foundation.[8] So, the term "dog", while it purports to represent a substance, fails to do so. Thus, the scientist of human nature recasts the term "dog" as a term for a mode, a term for a collection of qualities with no foundational uniting principle. The use of the general term "dog" is reconceived in the science of human nature to reflect *nothing more than* the mere associative tendencies of the human mind. This is the difference that sets apart Hume's own distinctions from those of the vulgar. Because, for example, the simple-complex distinction and the impression-idea distinction are more explanatory than those of the vulgar, the distinctions deployed by the science of human nature replace those of the vulgar and thereby come to represent the substance of the human mind. Thus, another way of putting the conclusion of the previous chapters is that while we ought to take Hume to be a Nominalist about the representation of modes, we must also accept the Ontological Interpretation of the representation of substances. Additionally, it is for this reason that we must understand the language of science as both committing us to the reality of its theoretical-explanatory posits, and to the recasting of that which is represented by the vulgar as substance as mere modes.

The thought here is this. What Hume's theory of general representation offers is a way of accounting for the meaning of general terms without the need to appeal to any real qualities or relations of the objects represented by such terms. General terms do not refer to universals, abstract entities, Platonic Forms, or robustly ontological relations between worldly items. The meaning of a general term consists *entirely* of a representation of the members of its revival set as resembling each other. At least, that is Hume's account of general terms *per se*. Consider, however, Hume's closing remark from 1.3.8 "Of the causes of belief".

> I must not conclude this subject without observing, that 'tis very difficult to talk of the operations of the mind with perfect propriety and exactness; because common language has seldom made any very nice distinctions among them, but has generally call'd by the same term all such as nearly resemble each other.
>
> T 1.3.8.15; SBN 105

Here we find Hume lamenting the paucity of conceptual resources afforded by common language, but what is revelatory about this passage is the explanation he gives for this paucity. The common language "has generally call'd by the same term all such as nearly resemble each other."[9] What is striking about that reason is that, as the Nominalist Interpretation would have it, such a procedure is precisely what Hume's theory of general representation limits *all* languages to doing. That is, the theory of general representation is supposed to account for all "abstract" representation via the resemblances among the ideas associated with a given general term. If that is right, though, Hume cannot very well object to common language on the grounds that it calls by the same term those ideas that resemble each other because this is how *all* language functions.

Recall, however, the distinction between the representations of substances and modes that we earlier saw Hume draw. After some parsing, we concluded that while representing both substances and modes makes use of the theory of general representation, representing the former, but not the latter, requires that the associations that form the revival set of the relevant general term operate via a "uniting principle" that serve as their "foundation." That is, the general representation that is used to represent a substance does not rely *merely* on resemblance, but also has some uniting principle as its foundation. And, of course, representing the essence, powers, and nature of substance is precisely what we saw Hume cast as the ultimate goal of the scientist of human nature. Thus, Hume's criticism of the common language: it relies on mere resemblance to structure its picture of the world, whereas the language of the scientist of human nature employs a more explanatory uniting principle as its foundation, thereby earning it greater ontological priority. The question that all of this prompts, then is this: how does the "uniting principle" that constitutes a general representation's being a representation of substance manifest itself? That is, what does this difference between the representation of substance and the representation of modes amount to in the use of a general representation?

To cut to the chase, the answer is that Hume's account of general representation does not specify *any* difference between these two kinds of representation (which is likely one reason that scholars have heretofore overlooked it). As far as the associative mechanisms of the theory of general representation are concerned, the representations of substances and modes are identical: we come to associate a word with a certain idea and stand disposed to recall other ideas that resemble this one as prompted. There is no role for the causal uniting principle to play here, and yet Hume tells us that it plays some role nonetheless. Since that role is not *internal* to the operations of a general term, then, it must be *external* to those operations. That is, for a general term to have uniting principle as its foundation is not for it to operate differently, but rather for its operations themselves to be treated differently, and specifically to be treated in such a way that they are taken to reflect a real fact about the ontological makeup of the world. Specifically, it is by *replacing* the language of the vulgar with the language of the science of human nature that the general representations of the latter but not the former come to represent *substances*.

By way of a brief first example, notice the pains that Hume goes to in the following passage to distinguish the ontology of the vulgar system from that of the scientist of human nature *via* the different distinctions that each draw in their respective languages.

> In order, therefore, to accommodate myself to their notions, I shall at first suppose; that there is only a single existence, which I shall call indifferently *object* or *perception*, according as it shall seem best to suit my purpose, understanding by both of them what any common man means by a hat, or shoe, or stone, or any other impression, convey'd to him by his senses. I shall be sure to give warning, when I return to a more philosophical way of speaking and thinking.
>
> T 1.4.2.31; SBN 202

Hume makes a point here of noting that in his subsequent discussion of the explanatory system of the vulgar, he will be employing the terms "object" and "perception" not according to the way that the scientist of human nature does, but rather according to the usage of the common man. It is important for Hume to note this change in idiom at this point in 1.4.2 precisely because that subsequent discussion is going to concern the *ontology* of the vulgar system: its attempt to posit a class of perceptions consisting of continued and distinct existences. The ontology of the vulgar is reflected in its use of general terms; the ontology of the science of human nature is reflected in its use of general terms; to substitute the latter for the former is thus to adopt a distinct set of ontological commitments.

Hume continues to engage with the language of the vulgar and its relation to the language of the science of human nature (or "the true philosophy") throughout 1.4 "Of skepticism and other systems of philosophy." Hume takes each of these systems to offer a different explanation of the apparent regularities discovered amongst our perceptions, mostly prominently the constancy and coherence of our perceptions. As such, these explanations are competitors for the title of "true philosophy" and the standard according to which they are judged is explanatory force.

Consider for example this passage from 1.4.2 "Of scepticism with regard to the senses," which will bring out several of these points. Here is Hume in the course of arguing that the explanation of the apparent constancy and coherence of our perceptions that the vulgar formulate is a product not of reason, but of the imagination.

> Accordingly we find, that all the conclusions, which the vulgar form on this head, are directly contrary to those, which are confirm'd by philosophy. For philosophy informs us, that every thing, which appears to the mind, is nothing but a perception, and is interrupted, and dependent on the mind; whereas the vulgar confound perceptions and objects, and attribute a distinct continu'd existence to the very things they feel or see.
>
> T 1.4.2.14; SBN 193

There are two important points to notice in just the first sentence here. The first of these is that Hume here makes very clear one of his differences from Berkeley: the conclusions of the vulgar are "directly contrary" to those of the scientist of human nature. Hume is not explicating common sense, but correcting it. Second, and more subtly, notice that Hume refers to the "*conclusions*, which the vulgar form on this head". Like that of the scientist of human nature, the language of the vulgar aims to *explain* something. The vulgar find themselves confronted with the apparent constancy and coherence of our perceptions, and form a picture—by using certain general terms—of the way the mind must be in order for that constancy and coherence to be a manifestation of its nature, essence, and powers. So, again, we can confirm that Hume is no mere inductivist about either the language of the vulgar or that of the scientist: the inductively-established regularities that the inductivist takes to be the end of scientific explanation are the very facts that both systems taken to be in need of explaining.

Moving on to the second sentence, we see Hume account for the difference in the conclusions of the vulgar and the scientist by appealing to the differences in the distinctions made by each. What the scientist represents as distinct, the vulgar represent as confounded, and in doing so "attribute a distinct continu'd existence to the very things they feel or see." That is, in accepting the language of the vulgar in which perceptions and objects are confounded, one thereby also accepts an ontology of perceptions in which a perception is "a distinct continu'd existence."

> This propension to bestow an identity on our resembling perceptions, produces the fiction of a continu'd existence; since that fiction, as well as the identity, is really false, as is acknowledg'd by all philosophers
>
> T 1.4.2.43; SBN 209

As we noticed in the previous section, taking a perception to be identical across time requires thinking of it as being of a certain kind, which in turn requires using a corresponding general term to represent it. Thus, to "bestow an identity on our resembling perceptions" the vulgar represent these perceptions using certain general terms.[10] In so doing, the vulgar represent a fictional world in which those perceptions continue to exist even when they are not perceived. They are committed to its ontology of continued existence because while all *philosophers* acknowledge the falsity of this explanatory picture, the vulgar themselves do not take it to be fictional, at all.

Of course, it is not just in their conception and ontology of body that the vulgar and the philosopher differ. Another revelatory section in this regard is 1.3.12 "Of the probability of causes," where Hume explains that while the vulgar understand probability as being a brute feature of the world, philosophers, who make finer distinctions, and therefore have more robust explanatory resources at their disposal, account for this apparent stochasticity by an appeal to its underlying causes.

> What I have said concerning the probability of chances can serve to no other purpose, than to assist us in explaining the probability of causes; since 'tis commonly allow'd by philosophers, that what the vulgar call chance is nothing but a secret and conceal'd cause.
>
> T 1.3.12.1; SBN 129

Notice that what the vulgar *call* chance, the philosopher takes to be nothing but a secret and concealed cause. As we noted earlier, the vulgar employ "common language" that "has seldom made any very nice distinctions among them, but has generally call'd by the same term all such as nearly resemble each other" (T 1.3.8.15; SBN 105). The vulgar notice the flip of a coin coming up sometimes heads, sometimes tails. They further notice that the uncertainty regarding this outcome resembles the uncertainty that attends the roll of a die. They thereby come to associate these phenomena (and others), and use the general term 'chance' to refer to them. Hume goes on to note that it is the crudeness of this distinction that accounts for the vulgar's taking the world itself to be stochastic in nature.

> *The* vulgar, who take things according to their first appearance, attribute the uncertainty of events to such an uncertainty in the causes, as makes them often fail of their usual influence, tho' they meet with no obstacle nor impediment in their operation. But philosophers observing, that almost in every part of nature there is contain'd a vast variety

> of springs and principles, which are hid, by reason of their minuteness or remoteness, find that 'tis at least possible the contrariety of events may not proceed from any contingency in the cause, but from the secret operation of contrary causes. This possibility is converted into certainty by farther observation, when they remark, that upon an exact scrutiny, a contrariety of effects always betrays a contrariety of causes, and proceeds from their mutual hindrance and opposition.
>
> T 1.3.12.5; SBN 132

The vulgar lack the conceptual resources to adequately explain not only the regularities of experience, but also any irregularities that arise. Limited in this way, the vulgar go so far as to the attribute these irregularities to the irregularity of causes themselves. Philosophers, by contrast, have significantly more robust conceptual resources and so can explain both the regularities of experience and its irregularity by appealing to the nature of substance underlying these phenomena. The philosopher thus recasts that which the vulgar take as explanatorily basic as in fact being the conglomerate effects of an underlying cause. The philosopher appeals to "a vast variety of springs and principles, which are hid" to replace the stochastic ontology of the vulgar with the more strictly causal ontology of science.

Of course, having *more* distinctions is not always better than having fewer. What matters is the explanatory force of the distinctions being drawn. So, consider the form that Hume's argument takes again the vulgar understanding of judgment, etc.

> This error consists in the vulgar division of the acts of the understanding, into *conception*, *judgment* and *reasoning*, and in the definitions we give of them. [. . .] What we may in general affirm concerning these three acts of the understanding is, that taking them in a proper light, they all resolve themselves into the first, and are nothing but particular ways of conceiving our objects. [. . .] This act of the mind has never yet been explain'd by any philosopher; and therefore I am at liberty to propose my hypothesis concerning it; which is, that 'tis only a strong and steady conception of any idea, and such as approaches in some measure to an immediate impression.
>
> T n20; SBN 96–97

Hume takes himself to be at liberty to propose his hypothesis, which replaces the trifold ontology of the vulgar with his own monistic ontology, because "this act of the mind has never been explain'd by any philosopher." That is, it is the explanatory force of a hypothesis that warrants the commitment to its attendant ontology. It is by replacing the conceptual distinctions of one system of explanation with those of another that we undertake this commitment. Whereas the vulgar take thought to consist of three distinct acts, the scientist of human nature casts all three of these activities as mere modes of a single substance: "they all resolve themselves into the first, and are nothing but particular ways of conceiving our objects."[11]

Finally, returning to 1.4, here is a passage in which Hume contrasts the ontology of the vulgar with that of the scientist of human nature by way of the "confounded" way of thinking of the former.

> *We* have a distinct idea of an object, that remains invariable and uninterrupted thro' a suppos'd variation of time; and this idea we call that of *identity* or *sameness*. We have also a distinct idea of several different objects existing in succession, and connected together by a close relation; and this to an accurate view affords as perfect a notion of *diversity*, as if there was no manner of relation among the objects. But tho' these two ideas of identity, and a succession of related objects be in themselves perfectly distinct, and even contrary, yet 'tis certain, that in our common way of thinking they are generally confounded with each other.
>
> T 1.4.6.6; SBN 253–254

> The scientist of human nature, of course, recognizes the diversity in the perceptions that constitute the self, and so adopts the ontology of the bundle theory. The common way of thinking runs this diversity together with what we call 'identity' and 'sameness', thereby confounding what the scientist keeps distinct, and thereby committing the vulgar to an ontology of a "soul, and self, and substance"
>
> T 1.4.6.6; SBN 254

To sum, then, because the language of the vulgar and the science of human nature each employ different general terms, they thereby picture the world as containing different kinds of substance. As we have seen, the theory of general representation itself does not distinguish between general terms that represent substances and those that do not, so there is no *intra*-linguistic way of representing the world as containing some particular substance rather than another. Thus, to accomplish this requires an *inter*-linguistic difference, namely, adopting one set of general terms in favor of another. Since representing a mode requires using a general term with no uniting principle as its foundation, and since we concluded that to use a general term *with* a uniting principle as its foundation is just to adopt a language that includes such a term, it follows that all successor languages cast the terms of their predecessor languages as modes. Finally, we have just now seen ample evidence that Hume advocates for the abandonment of the language of the vulgar on the grounds of the greater explanatory force of the language of the science of human nature.

As I have suggested, that explanatory force accrues only to representations that are a product of reason, and the final section of this chapter will concern precisely why that is. Specifically, it will explicate how Hume understands reason, its role in forming theoretical representations, and its mechanism for ensuring the reliability of these representations. It is to those arguments that I will now turn.

Reason, Memory, and Imagination

In the previous chapter, I presented evidence that Hume takes reason to play an ineliminable role in proper scientific theorizing, and as we will see, this role is so great that he takes a theoretical posit's being a product of faculty other than reason as sufficient for undermining the legitimacy of its use. I have also presented evidence that the kind of reasoning that Hume sees as occupying this role is not merely demonstrative reasoning, or the probable reasoning that produces the inductive generalizations that the science of human nature aims to explain, but also a kind of inference to the best explanation. It is because of the success of the simple-complex distinction in explaining the great novelty of human thought in the face of the relative paucity experiential content that we ought to accept a real distinction between simple and complex perceptions. It is because of the success of the impression-idea distinction in explaining the systematic connectedness of thought and experience that we ought to accept it, etc. What I have not done to this point, however, is to give much of an account of the faculty of reasoning itself, and my plan for this section is to do so.

Parts of the account that I will give here have already been foreshadowed, as have some of the particular bits of textual evidence. Additional support for it also comes holistically, from the benefits that accrue from adopting the interpretation of Hume's understanding of the science of human nature for which I have been arguing. Still, there is more work to be done, and that is the plan. With that, here are the most important theses regarding Hume's understanding of reason, which I will defend in what follows.

- Reason, like the imagination, is a recombinatory faculty, and thus produces novel representations.
- Reason, like memory but unlike the imagination, produces representations that are *reliable*.
- Unlike memory, reason's representations do not reliably track merely the past data of the senses, but also because of their explanatory force also reflect the truth about the substances underlying such experiences, and therefore can be used to track future experience as well.
- The representations that reason produces occupy a place on the spectrum of degrees of force and vivacity typically greater than imagination, and typically lower than memory.
- Memory, imagination, and reason—the three faculties that Hume includes in his theoretical apparatus—are constituted not by the degree of force and vivacity of their representations, but rather by two other criteria: whether they are reproductive (memory) or recombinatory (imagination and reason) and whether their function is to explain some phenomenon (reason) or not (imagination).

As this list of theses indicates, the plan for this section is to investigate the faculty of reason by way of contrasting it with the two faculties about which

Hume is more explicit: memory and imagination. As with so many things in Hume scholarship, this is itself a task that can seem straightforward enough at the outset, but very quickly becomes significantly more complicated. To see what has vexed scholars concerned with the distinction between memory and imagination, consider two passages. In the first, Hume presents an "evident" difference between the two faculties.

> There is another difference betwixt these two kinds of ideas, which is no less evident, namely that tho' neither the ideas of the memory nor imagination, neither the lively nor faint ideas can make their appearance in the mind, unless their correspondent impressions have gone before to prepare the way for them, yet the imagination is not restrain'd to the same order and form with the original impressions; while the memory is in a manner ty'd down in that respect, without any power of variation.
>
> T 1.1.3.2; SBN 9

Memory is "restrain'd to the same order and form with the original impressions [. . .] without any power of variation." Imagination is not restrained in this respect.[12] That is, memory is our faculty for *reproducing* the pictorial character of complex impressions in complex ideas whereas imagination is our faculty for *recombining* their constituent simple ideas into novel complexes.[13] Sounds good. The problem is that in a bit of text farther along in the *Treatise*, Hume appears to explicitly reject this evident difference as the criterion for what makes memory different from imagination.

> These faculties are as little distinguish'd from each other by the arrangement of their complex ideas. For tho' it be a peculiar property of the memory to preserve the original order and position of its ideas, while the imagination transposes and changes them, as it pleases; yet this difference is not sufficient to distinguish them in their operation, or make us know the one from the other; it being impossible to recal the past impressions, in order to compare them with our present ideas, and see whether their arrangement be exactly similar. Since therefore the memory is known, neither by the order of its *complex* ideas, nor the nature of its *simple* ones; it follows, that the difference betwixt it and the imagination lies in its superior force and vivacity.
>
> T 1.3.5.3; SBN 85

This is a strange passage for a number of reasons. To see this, consider a quick reconstruction of the argument Hume makes here.

(1) Memory merely reproduces impressions while the imagination recombines them.
(2) We cannot recall past impressions to compare with our present ideas.

(3) Therefore, we cannot "distinguish [memory and imagination] in their operations" by these characteristics. (1, 2)

Given that Hume clearly holds that we *can* know that the *faculty* of memory reproduces impressions whereas the *faculty* of imagination recombines them—he asserts that in the first premise—what is at issue in the second premise and conclusion can only be whether we can know that a *particular* idea reproduces some past impression or not. That is what Hume has to mean in concluding that we cannot distinguish memory from imagination "in their operations." Memory and imagination operate to reproduce and recombine, respectively, but we cannot *tell* which was operative in producing any given idea. What is striking about the passage, though, is that Hume's conclusion appears to be explicitly aimed not at what we can *tell* about *particular ideas*, but rather about the nature of these faculties themselves. What Hume appears to conclude is that the difference between the *faculty* of memory and the *faculty* of imagination is in "superior force and vivacity" (presumably the superior force and vivacity of the ideas produced by each). Insofar as it supports anything, Hume's argument here supports skepticism about whether a particular idea is a product of memory or imagination, but his conclusion is that these faculties differ only in the degree of force and vivacity of their product-representations. That is not good.

So, in the first passage Hume appears to weakly endorse a *prima facie* plausible criterion of the difference between memory and imagination.[14] In the second passage Hume appears to strongly endorse a *prima facie* highly implausible criterion of this difference, and for no good reason. Thus it is that scholars are left to puzzle out which if either criterion Hume actually endorses, and what to do with whichever criterion is left over. Here, for example, is a recent suggestion of Jeffrey McDonough's.

> Recall that Hume is concerned with sorting three different kinds of ideas: (a) accurate memories, (b) inaccurate memories, and (c) mere imaginings.[15] These three kinds of ideas could be intuitively sorted in either of two different ways:
>
> 1 (accurate memories and inaccurate memories) vs. (mere imaginations)
> 2 (accurate memories) vs. (inaccurate memories and mere imaginations)[16]
>
> Hume fails to resolve the tension between his two proffered criteria not because of a naïve conflation concerning how memories and imaginations are distinguished in practice with what in fact the distinction is between them, but rather from his attempting to mark two different, intuitive, distinctions between (a) accurate memories, (b) inaccurate memories, and (c) mere imaginings. His failure to reconcile the two criteria must not—implausibly—be supposed to result from careless oversight, but rather from his finding the phenomenal criterion compelling

when thinking broadly in terms of (1), and finding the constitutive criterion compelling when thinking broadly in terms of (2).[17]

McDonough's thought is that the difference of degrees of force and vivacity is a plausible enough way to distinguish memories (broadly construed) from imaginations, whereas reproduction vs. recombination is a plausible enough way to distinguish (accurate) memories from recombinations (broadly construed), and that the confusion of these different *explicanda* accounts for whatever blurring of these distinctions is to be found in Hume's text.

McDonough's interpretation would be compelling except for the fact that it still depends on attributing a rather serious although understandable confusion to Hume. Given the apparent conflicting texts, it may seem that some such gambit is unavoidable. I think that it is not. My own interpretation begins with the observation that we found a strikingly similar problem plaguing Hume's discussion of the impression-idea distinction. Hume appears to open the *Treatise* with the claim that the difference between impressions and ideas is in their respective degrees of force and vivacity. He then almost immediately concedes that some ideas have degrees of force and vivacity as great as that of impressions, which should be impossible. Our solution there was to understand Hume as using the degrees of force and vivacity of our perceptions (along with a few other indicators) to sort those perceptions into two resemblance classes, which paves the way for his investigation into the *reality* of the difference between these kinds of perception. The first step of this investigation comes in noting that all the members of the one class (ideas) are copies of the members of the other class (impressions). The next step comes in recasting this matter of fact about the two classes as marking a *real* difference between them, i.e., as marking an *ontological* difference in the substance of the mind that *explains* the aforementioned qualities of each (e.g., their different degrees of force and vivacity).

What I want to suggest regarding the distinction between memory and imagination is that the same dialectic is at work here. We use force and vivacity (along with a few other indicators) to pre-theoretically sort our perceptions into two resemblance classes, which sorting is then itself explained by a theoretical posit that marks an ontological difference in the constituents of the mind. Here the resemblance classes are roughly our more forceful and vivacious ideas and our less forceful and vivacious ones, and the theoretical posit is the existence of two faculties for producing these—the memory and imagination—which are defined by the way in which they structure complex ideas, and the existence and functioning of which explains, e.g., the difference of degree of force and vivacity of their respective products, their reliability in tracking the truth, etc. For example, because memory functions to reproduce, memories tend to resemble the impressions from which they are derived, and this resemblance causes these memories to retain a portion of the force and vivacity of those original impressions.[18] Likewise, memory's function of reproducing results in ideas that more reliably track past impressions than those that are the result of the imagination's arbitrary whims, etc.[19]

This reading of Hume's distinction between memory and imagination is useful in understanding Hume's general approach to the distinction and in rescuing some of the relevant texts. Another key piece of my strategy, though, is to see that many of the most troublesome texts occur in contexts in which Hume is concerned with the memory-imagination distinction only insofar as these representations *influence belief*. Since a belief, for Hume, is an idea with an appropriately high degree of force and vivacity, even if the difference between memory and imagination *is* the way in which each structure their product-representations, it would be natural to find him giving pride of place to the difference in their degrees of force and vivacity instead. So, in 1.1.3 "Of the ideas of the memory and imagination" where Hume is considering memory and imagination *per se*, the dialect outlined above appears more or less straightforwardly. By contrast, the more problematic passages emphasizing the role of force and vivacity come almost entirely in 1.3.5 "Of the impressions of the senses and memory" or thereabouts where Hume's focus is on the respective influences of memory and imagination on belief. So, rather than read Hume as being *confused* about which criterion to use to distinguish memory from imagination, on the reading I will offer, Hume is himself clear on the difference between these being constituted by the structure of their product-representations. Since he is also committed, though, to the thesis that this difference is only indirectly responsible for the difference in the way these two faculties influence belief, in the context of discussing that, he treats force and vivacity as criterial instead.

With those two interpretive principles articulated, we can begin with the paragraph immediately preceding the first one above (the one where Hume appears to endorse distinguishing memory from imagination according to whether an idea is a reproduction or recombination).

> We *find by experience*, that when any impression has been present with the mind, it again makes its appearance there as an idea; and this it may do after two different ways: either when in its new appearance it retains a considerable degree of force and vivacity, and is somewhat intermediate betwixt an impression and an idea; or when it intirely loses that vivacity, and is a perfect idea. The faculty, by which we repeat our impressions in the first manner, is *called* the MEMORY, and the other the IMAGINATION. 'Tis *evident at first sight*, that the ideas of the memory are much more lively and strong than those of the imagination, and that the former faculty paints its objects in more distinct colours, than any which are employ'd by the latter. [. . .] Here then is a *sensible difference* betwixt one species of ideas and another.
>
> T 1.1.3.1; SBN 8–9, emphasis mine

There are a number of interesting things to notice about this paragraph. First, notice that this passage *immediately precedes* the one in which Hume differentiates the memory from the imagination via the difference between reproducing and recombining, and yet appears like the second passage above, to cast this distinction in terms of degree of force and vivacity. So, the suggestion that Hume just got *confused* about these two ways of drawing the distinction, for whatever reason, is highly implausible. He clearly had both in mind, as distinct, at one time.

Second, note that Hume cites this way of drawing the distinction, via degrees of force and vivacity, as *found by experience* and as made via a "*sensible* difference between one species of ideas and another." And that rather than write that the perceptions so distinguished *are* memories and imaginings, he notes that they are *called* "memory" and "imagination". That is, just as with impressions and ideas, we sort our perceptions into two resemblance classes according to their introspectively observable similarities (degree of force and vivacity), and then affix the *words* "impressions" and "ideas" (as per the theory of general representation), only *then* to investigate the real difference between these, Hume here repeats that exact procedure. Sort our perceptions according to their introspectively available properties, affix a general term to each class, and then investigate the real difference between them. Hume presents the results in the following paragraph, which we saw earlier.

> There is another difference betwixt these two kinds of ideas, which is no less evident, namely that tho' neither the ideas of the memory nor imagination, neither the lively nor faint ideas can make their appearance in the mind, unless their correspondent impressions have gone before to prepare the way for them, yet the imagination is not restrain'd to the same order and form with the original impressions; while the memory is in a manner ty'd down in that respect, without any power of variation.
>
> T 1.1.3.2; SBN 9

Which, after a few examples, continues,

> In short, this principle [that the chief exercise of the memory preserves the order and position of simple ideas] is supported by such a number of common and vulgar phenomena, that we may spare ourselves the trouble of insisting on it any farther.
>
> The same evidence follows us in our second principle, *of the liberty of the imagination to transpose and change its ideas.*
>
> T 1.1.3.3–4; SBN 9–10

Notice that by these concluding remarks Hume has changed the status of the observed differences of behavior of "memories" and "imaginings" from evident differences to *principles*. That is, after sorting our ideas according to their introspectively available and pre-theoretical qualities, Hume quickly finds a more ontologically robust difference between these two classes, and subsequently codifies and reifies this difference. By the close of this section, it is no longer a mere matter of fact that memories and imaginings are reproductions and recombinations, respectively, but is a theoretical-explanatory principle that they are. That is the real difference between them that explains the introspective qualities that allowed for us to pre-theoretically track this more basic ontological difference.

What, then, of the passage from later in the *Treatise* where Hume explicitly rejects using this principle as a criterion for distinguishing memories from imagination? First, note that we were able to generate a conflict similar to the one that that later passage caused with the passage above, which was easily ameliorated. Second, recall that while the later passage is more explicit about using force and vivacity as the criterion for distinguishing memories from imaginings, it also presents a very bad argument for doing so, and so if we have to choose a passage to reject, it ought to be the one that is intrinsically worse. Third, there is other evidence from the very same section in which the later passage appears that Hume takes reproduction and recombinability to be the criterion of difference between memory and imagination.

> We are frequently in doubt concerning the ideas of memory, as they become very weak and feeble; and are at a loss to determine whether any image proceeds from the fancy or the memory, when it is not drawn in lively colours as distinguish that latter faculty
>
> T 1.3.5.5; SBN 85

> And as an idea of the memory, by losing its force and vivacity, may degenerate to such a degree, as to be taken for an idea of the imagination; so on the other hand an idea of the imagination may acquire such a force and vivacity, as to pass for an idea of the memory, and counterfeit its effects on the belief and judgement.
>
> T 1.3.5.6; SBN 86

Just as in the case of impressions and ideas, degree of force and vivacity alone is an unreliable indicator of the nature of a perception. That is, just as some ideas can be as forceful and vivacious as some impressions (as in madness or fever), so can some memories be as weak and feeble as some imaginings. For that to be possible, however, it must be that what makes an idea a memory or an imagining is something other than its degree of force and vivacity.[20]

Finally, and most importantly, it is crucial to understand the context in which these two considerations of this distinction occur. The pair of passages from the earlier section of the *Treatise* occurs where Hume's attention is squarely focused on what makes memory and imagination different.

There he mentions both force and vivacity and production and reproduction, and appears to use only the latter as criterial. In the later section, Hume is considering that distinction only *insofar as it makes a difference to the causes of our beliefs*. That is, Hume's focus in the bad-argument passage is not on marking the difference between memory and imagination *per se*, but rather in understanding why it is that the former but not the latter faculty *results in belief*. In fact, in his concluding remarks on that score, Hume takes differences in degree of the force and vivacity to mark not the difference between memories and imaginings, but rather that between imaginings on the one hand and *impressions or memories* on the other.

> Thus it appears, that the *belief* or *assent*, which always attends to the memory and senses, is nothing but the vivacity of those perceptions they present; and that this alone distinguishes them from the imagination. To believe is in this case to feel an immediate impression of the senses, or a repetition of that impression in the memory.
>
> T 1.3.5.7; SBN 86

Hume's ultimate thesis in this later section is not that the difference between a memory and an imagining *per se* is its degree of force and vivacity, but rather that *beliefs*, which always *attend* impressions and memories, are distinguished from imaginings, or mere conceptions, by their degree of force and vivacity.

That Hume's conclusion is about what distinguishes a belief from an imagining also makes the bad argument from the later passage a better one. Recall that that argument seemed to proceed from the lemma that the only way to distinguish a memory from an imagining is via is degree of force and vivacity to the conclusion that the difference between the faculty of memory and the faculty of imagination is in the degree of force and vivacity of its product-representations. That bad inference was made even worse by the fact that in the very same passage Hume acknowledged that the faculty of memory and the faculty of imagination differ insofar as the former is reproductive whereas the latter is recombinatory. What we can now see was happening there is that Hume's conclusion is really that *with respect to belief formation* the only difference between memory and imagination is the degree of force and vivacity of their product-representations. That conclusion does follow from the fact that we cannot compare past impressions with current ideas because it follows that such a comparison is not *the basis of belief*.

This modified conclusion is also compatible with the real difference between these faculties being that memory is reproductive whereas imagination is recombinatory because the *real* difference between them is only tangentially relevant to the difference between them with respect to belief. The real difference *explains* the difference in degrees of force and vivacity, and thereby explains the different influences each faculty has on belief formation. That is why Hume cites the real difference between memory and imagination at the start of the passage! Here is the passage again.

> These faculties are as little distinguish'd from each other by the arrangement of their complex ideas. For tho' it be a peculiar property of the memory to preserve the original order and position of its ideas, while the imagination transposes and changes them, as it pleases; yet this difference is not sufficient to distinguish them in their operation, or make us know the one from the other; it being impossible to recal the past impressions, in order to compare them with our present ideas, and see whether their arrangement be exactly similar. Since therefore the memory is known, neither by the order of its *complex* ideas, nor the nature of its *simple* ones; it follows, that the difference betwixt it and the imagination lies in its superior force and vivacity.
>
> T 1.3.5.3; SBN 85

The dialectic of this passage should be read as follows. The real difference between memory and imagination is that the former is reproductive and the latter is recombinatory. However, that difference does not directly influence belief formation, "is not sufficient to distinguish them in their operation," because we cannot recall past impressions to compare them with present ones. So, insofar as belief formation is concerned, the relevant difference between these two faculties is the degree of force and vivacity of their product-representations. If such a reading is on target, then whatever the evidential force of the later passage is in establishing that degree of force and vivacity is criterial, what it is a criterion for is not distinguishing memory from imagination, but for distinguishing *belief* from conception, which is a much more plausible and familiar thesis of Hume's.

Thus, I conclude that the difference between a memory and an imagining for Hume is that the former is the product of a faculty the function of which is to reproduce the order and position of our impressions whereas the latter is a product of a faculty that recombines simple ideas into complex ideas with novel representational content. That conclusion, now leaves me with a bit of a puzzle, as I began this section by claiming that *reason* differs from memory insofar as it recombines the order and position of our impressions to create representations with novel content, but also that reason is *not* the same faculty as the imagination. Before presenting and defending the solution to that little puzzle, it is worth noting here that this entire claim—that Hume considers reason and imagination distinct faculties, and that reason produces novel content of its own—is rather controversial. For example, Garrett extracts the following argument from the *Treatise* to support Hume's rejection of a Cartesian intellect, "a higher faculty whose representations are radically nonimagistic ideas".

1. [A]ll our ideas are copy'd from our impressions.
2. All impressions are clear and precise.
3. The ideas . . . must be of the same nature [as impressions], and can never, but from our fault, contain anything . . . dark and intricate. (from 1 and 2)

4. [There are no] spiritual and refined perceptions . . . [that] . . . must be comprehended by a pure intellectual view, of which the superior faculties of the soul are along capable. (from 3)[21]

Without delving into the details very deeply, that strikes me as a perfectly good argument to attribute to Hume, and the conclusion seems right on target. There are no nonimagistic ideas, and we have no "intellectual" faculty whose function it is to produce such ideas. However, Garrett goes on to infer from this conclusion that,

> It is because of Hume's confidence in this conclusion—equivalent to the rejection of the Cartesian intellect—that he can feel free to make an exhaustive distinction of idea-forming faculties into *memory* and *imagination*, faculties who representations differ *not* in fundamental character or content but only in causal history and in degree of "force and liveliness" or "vivacity".[22]

While Garrett is certainly correct that Hume rejects the Cartesian intellect, it does not follow from this that the memory and imagination are the only faculties for recombining ideas.[23] Garrett does not reveal much about his support for this claim, but his thinking here might be as follows. If all thought is imagistic, and ideas are all derived from impressions, then the only faculty that we have for producing ideas or thoughts is one that derives the content of its representations from impressions. Memory and imagination are both species of such a faculty, the constitutive difference between them being a difference in the degree of force and vivacity of the representations they each produce (although Garrett does also mention their causal history). If one supposes that memories are those representations with a degree of force and vivacity above a certain threshold, and imaginings are those below that threshold, then it does seem to follow that there is no place for any representative faculty other than these two.

The problem with this line of reasoning is that it neglects a number of ways of carving out distinct faculties. For example, even if we use only degrees of force and vivacity to distinguish faculties, there is nothing preventing us from distinguishing more than *two* such faculties. For example, one might employ as many thresholds for defining faculties as there are distinct degrees of force and vivacity like so:

imagination $< x < y < \ldots n <$ memory

One *could* cast reason, although one ought not to, as a distinct third faculty that produces representations with a degree of force and vivacity intermediate between memory and imagination. Alternatively, once one abandons degree of force and vivacity as the constitutive the difference between memories and imaginings, even more possibilities become available. For

Table 3.1

	Caused by a resembling impression	*Not Caused by a resembling impression*
Degree of Force and Vivacity > x	Memory	Faculty3
Degree of Force and Vivacity < x	Faculty4	Imagination

example, even using just the criteria that Garrett indicates, one could carve out faculties as follows.

On this way of carving the conceptual space, the difference between memory and imagination is both their differing degrees of force and vivacity and their differing causal histories, but these two orthogonal criteria also make possible two other faculties. Faculty$_3$ would share with memory the production of ideas with a high degree of force and vivacity, but would differ in its causal history. Faculty$_4$ would produce ideas that were derived from resembling impressions, but which would have a comparatively low degree of force and vivacity. Using other criteria, once could demarcate still more faculties.

The point here is not to argue that Hume countenanced either faculty$_3$ or faculty$_4$, or that he used both degree of force and vivacity and causal history as criteria for demarcating faculties. Rather, the point is that Garrett's conclusion does not follow from the premises that he presents above. Hume can and does reject the Cartesian intellect without limiting himself to only two faculties. He is perfectly free to posit some third faculty, distinct from both the memory and imagination. All that the rejection of the Cartesian intellect accomplishes is to eliminate the possibility of a faculty for producing nonimagistic representations. It does not thereby also eliminate the possibility of there being faculties that differ in the imagistic representations that they produce in ways that are different from the ways that the memory and imagination differ from each other.[24]

Of course, as I have indicated, I take reason to be just such a faculty. Here is an instance of Hume contrasting these three faculties with one another.

> When I oppose the imagination to the memory, I mean the faculty, by which we form our fainter ideas. When I oppose it to reason, I mean the same faculty, excluding only our demonstrative and probable reasonings.
> T n22; SBN 117

Now, this passage is a little thorny as it makes it sound both like the constitutive difference between imagination and memory is the degree of force and vivacity of their product-representations, and like reason is a species of imagination. As earlier, though, the context of this remark makes clear why Hume is writing this way, and why we should take these claims neither literally nor to apply to the difference between these faculties *simpliciter*. Specifically, Hume opens the paragraph of which this sentence is a part by noting that,

> In general we may observe, that as our assent to all probable reasonings is founded on the vivacity of ideas, it resembles many of those whimsies and prejudices, which are rejected under the opprobrious character of being the offspring of the imagination.
>
> Tn22; SBN 117

As this is a note to one of the sections on the causes of *belief*, Hume is here paying particular attention to the fact that the products of reason, just like the products of the imagination (and memory and senses), only become *beliefs* by achieving a high enough degree of force and vivacity. So, once again, insofar as it is relevant to belief, the only distinction between the faculties that matters is their degree of force and vivacity. In fact, in the text to which this note is appended, Hume has just finished *lamenting* that some ideas of the imagination can, through habituation, have degrees of force and vivacity exceeding those of reason, although not those of memory. So, when Hume distinguishes memory from imagination and imagination from reason in the passage above, with his focus once again squarely on belief, he emphasizes the degrees of force and vivacity of their product-representations. In doing so, the high degree of force and vivacity of memory warrants its own category; the lower degrees of force and vivacity of reason and imagination then get lumped together.

Still, it is worth noting here how Hume distinguishes imagination and reason within their own category: reason is our faculty for conducting demonstrative and probable reasonings. That is noteworthy insofar as it marks the novel representations of reason via the processes from which they result. If we keep in mind that the *real* difference between memory on the one hand and imagination and reason on the other is that the former reproduces the order and position of impressions whereas the latter recombines simple impressions into new orders and positions, then the difference between imagination and reason, the two recombinatory faculties, becomes the process whereby their recombinations are produced. As we will see in a moment, the constitutive difference in those processes is that while the imagination is free to recombine representations *arbitrarily*, according to custom, habit, or whim, reason's recombinatory activities serve a particular function. Just as memory is restrained by its function of reproducing the order and positions of past impressions, reason is similarly restrained by its function of *explaining* empirically discovered regularities of perception.[25]

That is what I take the constitutive difference between reason and the imagination to be. That *constitutive* difference also leads to other merely symptomatic differences between reason and the imagination as well. For example, the products of reason typically but not always have a higher degree of force and vivacity than do those of the imagination. Also, the products of reason are more epistemically reliable than those of the imagination.[26] Finally, because of their explanatory force, we are ontologically committed to the objects of the product-representations of reason, whereas there is nothing ontologically committing about imaginings. I will address each of these features of the differences in turn.

The first item on my current agenda, then, is to show that reason is a recombinatory faculty the function of which is to explain empirically discovered regularities of perception. While Hume is not explicit on this point, many of the examples that we had occasion to consider in previous chapters make it clear that he must be thinking along these lines. Here briefly are a few of those passages.

> This mistake [that the senses represent as minute and uncompounded what is really great and compos'd of a vast number of parts] we are not sensible of; but taking the impressions of those minute parts, which appear to the senses to be equal or nearly equal to the objects, and finding by reason, that there are other objects vastly more minute, we too hastily conclude, that these are inferior to any idea of our imagination or impression of our senses.
>
> T 1.2.1.5; SBN 28

The senses represent something as simple, but reason discovers it to be—forms a more accurate representation of it as being—complex. We do not, at least typically, represent the world's minutiae via the senses, but rather reason reveals to us that what the senses present cannot be taken at face value, and combines with what the senses have provided to formulate a more adequate explanation that phenomenon.

> 'Tis commonly allow'd by philosophers, that all bodies, which discover themselves to the eye, appear as if painted on a plane surface, and that their different degrees of remoteness from ourselves are discover'd more by reason than by the senses.
>
> T 1.2.5.8; SBN 56

What reason does in this case is the same as what it does in the case above: it both discovers that what the senses provide requires further explanation, and contributes to forming a theoretical representation of an entirely new kind of object. Here is another similar example.

> When you tell me of the thousandth and ten thousandth part of a grain of sand, I have a distinct idea of these numbers and of their different proportions; but the images, which I form in my mind to represent the things themselves, are nothing different from each other, nor inferior to that image, by which I represent the grain of sand itself, which is suppos'd so vastly to

> exceed them. [. . .] But whatever we may imagine of the thing, the idea of a grain of sand is not distinguishable, nor separable into twenty, much less into a thousand, ten thousand, or an infinite number of different ideas.
>
> T 1.2.1.3; SBN 27

Hume grants that we can form an idea of something a thousandth the size of a grain of sand—"I have a distinct idea of these numbers and of their different proportions"—but also holds that it is not the senses alone that produce that idea—"the images, which I form in my mind to represent the things themselves, are nothing different from each other." The distinct idea that we form of these numbers and their different proportions is that formed by considering the philosophical relation of "proportions in quantity or number" (T 1.1.5.6; SBN 14–15), and "except in very short numbers, or very limited portions of extension" (T 1.3.1.3; SBN 70) these depend on "a chain of reasoning" (T 1.3.1.5; SBN 71). So, here too we find reason (in this case, demonstrable rather than probable) playing a central role in moving beyond what is provided by the senses.

What these three passages, in addition to others, show is that reason is a recombinatory faculty the function of which is to explain the regularities found in experience for which the senses alone cannot account. They do not show that this is what *constitutes* reason, but I will return to that stronger claim at the close of this section, after we have considered some of the other differences between reason, memory, and imagination. The next such difference on our list was that the products of reason typically but not always have a higher degree of force and vivacity than do those of the imagination. Here is a representative passage on that score.

> And indeed such a fiction [of resembling and contiguous objects] is founded on so little reason, that nothing but pure *caprice* can determine the mind to form it; and that principle being fluctuating and uncertain, 'tis impossible it can ever operate with any considerable degree of force and constancy.
>
> T 1.3.9.6; SBN 109

Notice first that reason is here contrasted with "pure caprice." That is, whereas the imagination is free to recombine ideas according to its whims, reason operates within certain constraints. As I have suggested, it functions to explain. As such, the products of the imagination are "fluctuating and uncertain"—you never know *what* the imagination might produce—and so do not exhibit the steadiness or persistence that is required to establish the customs and habits necessary to impart high degrees of force and vivacity to them. By contrast, the ideas produced by reason, because they are formed on the basis of explanatory principles, would be more steady and firm, and would have an attendant high degree of force and vivacity.

Of course, some representations of the imagination are so firmly ingrained that their degree of force and vivacity does rise above that of reason.

> All those opinions and notions of things, to which we have been accustom'd from our infancy, take such deep root, that 'tis impossible for us, by all the powers of reason and experience, to eradicate them; and this habit not only approaches in its influence, but even on many occasions prevails over that which arises from the constant and inseparable union of causes and effects.
>
> T 1.3.9.17; SBN 116

While the representations produced by reason accrue a high degree of force and vivacity due to the "constant and inseparable union of cause and effect"—recall that reason functions to accurately representing *substance*, which is the common cause of a set of associated qualities—there are some representations of the imagination that can obtain a still higher degree simply in virtue of their longstanding place in our cognitive repertoire. That makes sense: degrees of force and vivacity correlate to the steadiness of a perception. One way of being steady is through mere custom and habit. Another way is through a kind of explanatory robustness: once a certain hypothesis is taken to explain a certain set of phenomena, the more each repetition of those phenomena comes to signal that hypothesis and establish a firm and steady place for *it* in the mind.

Thus it is that the high degree of force and vivacity that the representations of reason accrue compared to those of the imagination corresponds to their higher degree of epistemic reliability. It is because these representations are epistemically reliable that they have a higher degree of force and vivacity, even if the unreliable representations of the imagination do sometimes rise to that level.

> As liars, by the frequent repetition of their lies, come at last to remember them; so the judgment, or rather the imagination, by the like means may have ideas so strongly imprinted on it, and conceive them in so full a light, that they may operate upon the mind in the same manner with those, which the senses, memory, or reason present to us.
>
> T 1.3.9.19; SBN 117

Notice that Hume go out of his way to differentiate the imagination from the more reliable faculties: senses, memory, and *reason*. That is more evidence that reason is not identical to the imagination. This passage continues,

> But as education is an artificial and not a natural cause, and as its maxims are frequently contrary to reason, and even to themselves in different times and places, it is never upon that account recogniz'd by philosophers
>
> T 1.3.9.19

In this section, Hume casts education as the habituation of the imagination. He here not only once again contrasts such a trained imagination with reason as *de facto* natural and unnatural faculties, but also notes that each operates on contrasting "maxims," i.e., normative principles, and that it is only those of reason that ought to be endorsed (at least in part because they form a more *coherent* set of imperatives). That is, reason is not only more natural than education, a species of the imagination, but it is also more epistemically reliable. Here is Hume explicitly endorsing a very strong version of that thesis.

> Our reason must be consider'd as a kind of cause, of which truth is the natural effect; but such-a-one as by the irruption of other causes, and by the inconstancy of our mental powers, may frequently be prevented.
>
> T 1.4.1.1, SBN 179

Truth is the natural effect of reason, and it is only when other faculties interrupt reason's course that we are led astray. That is, reason is as epistemically reliable as Hume's system allows for anything to be—it is a cause the effect of which is truth—and we only ever fall into error when some other faculty interrupts reason's otherwise reliable process. Here is Hume repeating precisely the same evaluation later.

> For if we assent to every trivial suggestion of the fancy; beside that these suggestions are often contrary to each other; they lead us into such errors, absurdities, and obscurities, that we must at last become asham'd of our credulity. Nothing is more dangerous to reason than the flights of the imagination, and nothing has been the occasion of more mistakes among philosophers.
>
> T 1.4.7.6; SBN 267

The fancy, or imagination, produces representations that, "are often contrary to each other," and which result in, "such errors, absurdities, and obscurities, that we must at last become asham'd of our credulity." All of this is the most dangerous threat to *reason* insofar as it results in widespread and pernicious mistakes. What this shows again is that imagination and reason are two *distinct* faculties that work at cross-purposes insofar as imagination produces representations that are epistemically unreliable whereas reason functions to track the truth.

Here is one final passage to support that thesis.

> Now upon that supposition, 'tis a false opinion that any of our objects, or perceptions, are identically the same after an interruption; and consequently the opinion of their identity can never arise from reason, but must arise from the imagination.
>
> T 1.4.2.43; SBN 209

Hume concludes that the supposition at hand is a product of the imagination rather than reason solely on the grounds that it is false. That strongly implies that Hume takes there to be a close link between the reliability of reason in tracking the truth on the one hand, and the unreliability of imagination to do the same on the other.[27]

Moving on to the last of the theses regarding reason outlined above, the complement of the superior epistemic reliability of reason over that of imagination is that it is the representations of reason, rather than imagination, to which we ought to be ontologically committed. In 1.3.9 "Of the effects of other relations and other habits", Hume is concerned to contrast reason's discovery of relations of cause and effect with the imagination's associations of representations via relations of contiguity and resemblance. It is unsurprising to find him concluding that,

> the relation of cause and effect is requisite to perswade us of any real existence.
>
> T 1.3.9.6; SBN 109

Meanwhile, the imagination's association of resembling and contiguous ideas proves itself distinctly unreliable, and therefore unworthy of ontological commitment.

> And as this imperfection is very sensible in every single instance, it still increases by experience and observation, when we compare the several instances we may remember, and form a *general rule* against the reposing any assurance in those momentary glimpses of light, which arise in the imagination from a feign'd resemblance and contiguity.
>
> T 1.3.9.6; SBN 110

While this section is primarily concerned with continuing Hume's investigation of the *causes* of belief, notice that Hume here explicitly notes that in addition to the products of the imagination failing to *de facto* produce belief, they also, *de jure*, *ought not to* result in belief. We form a *rule* against committing ourselves to the existence of the objects represented by products of the imagination. Those representations, unlike the representations formed by reason, are not accurate representations of what really exists.

If these the preceding observations of the various contrasts that Hume draws between reason and imagination are correct, then while we have not yet determined what *constitutes* the difference between reason and imagination, we ought to be in a good position to do so. That is, we can now engage in a little bit the science of human nature of our own. Here again are the empirically discovered regularities in need of explanation.

- Reason and imagination both contrast with memory insofar as they are recombinatory rather than merely reproductive.
- Reason functions to *explain* empirically discovered regularities of perception, whereas imagination does not.

- The products of reason typically but not always have a higher degree of force and vivacity than do those of the imagination.
- The products of reason are more epistemically reliable than those of the imagination.
- We are ontologically committed to the objects of the product-representations of reasons, whereas there is nothing ontologically committing about imaginings.

My suggestion is if we take the reason to *be* a recombinatory faculty that functions to explain, i.e., take the first two matters of fact to mark off a real difference in cognitive operations, then we get a robust explanation the remaining three facts. It is because reason is such a faculty that its representations typically but not always have a high degree of force and vivacity, and that they are epistemically reliable, and that they are ontologically committing. Distinguishing reason in any other way does not have the same kind of explanatory power. For example, while it is a core tenet of scientific realism that explanatory power yields epistemic reliability and ontological commitment, reversing that order of explanation makes little sense. Likewise, it is not difficult to see how functioning to explain could yield a representation with a high degree of force and vivacity, but it is much more difficult to understand how merely having a high degree of force and vivacity could make an idea explanatory.

If that inference to the best explanation goes through, then we have arrived at our full characterization of reason. Reason is a recombinatory faculty that functions to explain the empirical regularities discovered in experience. As we know from previous chapters, its recombinations are the product of taking an observable phenomenon as a model, and specifying determinate ways that the object of its product-representation differs from and resembles that model. Its explanations come in the form of positing the existence of such theoretical entities by adopting a language in which the general terms for such objects replace the general terms of its predecessor language (and thereby cast the objects of that predecessor language as mere modes).

Before moving on, it is worth pausing to consider a few objections to this interpretation of reason. Consider, then, the following representative passage from the *Enquiry*.

> When a child has felt the sensation of pain from touching the flame of a candle, he will be careful not to put his hand near any candle; but will expect a similar effect from a cause, which is similar in its sensible qualities and appearance. If you assert, therefore, that the understanding of the child is led into this conclusion by any process of argument or ratiocination, I may justly require you to produce that argument; nor have you any pretence to refuse so equitable a demand. You cannot say, that the argument is abstruse, and may possibly escape your enquiry; since you confess, that it is obvious to the capacity of a mere infant. If you hesitate, therefore, a moment, or if, after reflection, you produce

> any intricate or profound argument, you, in a manner, give up the question, and confess, that it is not reasoning which engages us to suppose the past resembling the future, and to expect similar effects from causes, which are, to appearance, similar.
>
> EHU 4.23; SBN 39

On one reading of this passage, as well as others like it, Hume's thesis here is that it is not reasoning that leads to the connection between cause and effect, thus paving the way for the conclusion that is *merely* habit that does so. Notice, though, that Hume's conclusion here is that, "it is not reasoning which engages us to suppose *the past resembling the future*, and to *expect similar effects from causes*." That is, Hume's conclusion here is that it is not reason that underlies induction itself. As Garrett has argued, what this means is only that there is no argument the conclusion of which is that the past will resemble the future (the Uniformity Principle).[28] It does not follow from this thesis alone, though, that probable reasoning itself is illegitimate, but only that such reasoning rests on a thesis that is not *in turn* supported by reason. There is, of course, an abundance of evidence that Hume takes probable reasoning, reasoning from cause to effect and vice versa, to be perfectly legitimate, for example, the existence of 1.3.15 "Rules by which to judge of causes and effects."

Another set of passages that might cast doubt onto the understanding of reason that I have proposed are those in which Hume writes that, "reason alone can never give rise to any original idea" (T 1.3.14.5; SBN 157). One way of understanding that claim is as implying that reason is not itself a recombinatory faculty, but only a faculty for transitioning from one (previously formed) idea to another.[29] That is, one might take such passages to imply that reason is an inferential faculty that *merely* brings out relations of ideas (demonstrative reasoning) or discovers causal relations between distinct ideas (probable reasoning). I have argued that we ought to understand this inferential faculty more broadly as also including inference to the best explanation wherein reason, seeking an explanation for some inductively-established universal regularity recombines ideas to produce theoretical representations with novel content. Thus, I need to show that the kinds of passages mentioned above are compatible with that interpretation. Here, then, is the complete passage from which the above quotation is taken, which occurs in the course of Hume's consideration of the ideas of necessary connection, power, etc.

> *I believe* the most general and most popular explication of this matter, is to say, that finding from experience, that there are several new productions in matter, such as the motions and variations of body, and concluding that there must somewhere be a power capable of producing them, we arrive at last by this reasoning at the idea of power and efficacy. But to be convinc'd that this explication is more popular than philosophical, we need but reflect on two very obvious principles. *First*,

> That reason alone can never give rise to any original idea, and *secondly*, that reason, as distinguish'd from experience, can never make us conclude, that a cause or productive quality is absolutely requisite to every beginning of existence. Both these considerations have been sufficiently explain'd; and therefore shall not at present be any farther insisted on.
>
> T 1.3.14.5, SBN 157

It is important to note that Hume's target here is the thesis that we have an idea of power and efficacy that is *entirely unlike* any other idea insofar as it is not itself and *image*, and the attendant suggestion that such an idea might be the product of reason *alone*. These theses are both ones that we have previously seen Hume reject, and which are entirely compatible with holding that reason is a recombinatory faculty. What Hume argues here is only that reason cannot produce ideas that are different *in kind* from those derived from impressions. In this sense, the imagination, "can never give rise to any *original* idea," either. Neither the imagination nor reason produce novel ideas *ex nihilo*, but what novel content they do produce is a product of recombining the deliverances of the senses. Thus, Hume continues,

> *I shall* only infer from them, that since reason can never give rise to the idea of efficacy, that idea must be deriv'd from experience, and from some particular instances of this efficacy, which make their passage into the mind by the common channels of sensation or reflection. Ideas always represent their objects or impressions; and *vice versa*, there are some objects necessary to give rise to every idea.
>
> T 1.3.14.6, SBN 157–158

When Hume writes here that the idea of efficacy, or ideas more generally, must be derived from experience, it is should be obvious that he does not mean that it must be *copied* from some impressions full stop. Again, we already know that the imagination recombines ideas, and that even when a complex idea is a memory, it need not *exactly* reproduce the complex impression from which it is derived. So, what Hume clearly means here is just that ideas must be copies *or recombinations* of impressions or experience. That thesis does not rule out reason's producing an idea that is a recombination of ideas themselves derived from experience. It only eliminates the possibility of reason producing ideas that bear no relation to experience whatsoever, or of are of an entirely different kind than impressions. (That reason cannot do *that* is the portion of Garrett's anti-Cartesian thesis that I endorsed earlier.)

Finally, consider the following passage in which Hume appears to claim that causal "reasoning" does not actually involve reasoning at all, but is entirely a product of the imagination.

> Reason can never shew us the connexion of one object with another, tho' aided by experience, and the observation of their constant

> conjunction in all past instances. When the mind, therefore, passes from the idea or impression of one object to the idea or belief of another, it is not determin'd by reason, but by certain principles, which associate together the ideas of these objects, and unite them in the imagination. Had ideas no more union in the fancy than objects seem to have to the understanding, we cou'd never draw any inference from causes to effects, nor repose belief in any matter of fact. The inference, therefore, depends solely on the union of ideas.
>
> T 1.3.6.12; SBN 92

Of course, this passage comes from the close of Hume's famous discussion of the Uniformity Principle: its origin, use, and/or justification. It is not necessary to delve into the details of that discussion just now, because this general context is enough to shed light on Hume's actual intention in this passage. What Hume is claiming here is not that *probable* reasoning can never show us the connection of one object with another, but rather that *demonstrative* reasoning cannot do so. He introduces this discussion a few paragraphs earlier by noting that,

> 'Twere easy for me to shew the weakness of this reasoning, were I willing to make use of those observations, I have already made, that the idea of *production* is the same with that of *causation*, and that no existence certainly and demonstratively implies a power in any other object
>
> T 1.3.6.9; SBN 90

Hume declines to make use of the particular argument mentioned here, but the target of the argument that he does use is the same: the claim that some existence "*certainly and demonstratively* implies a power in any other object." Thus, Hume's conclusion here is not that all reasoning reduces to imagination, but rather that *demonstrative* reasoning does not reveal any necessary connections between distinct objects, and that any (non-necessary) connections that are discovered among distinct objects can only be the product of the faculties that Hume has already delineated including of course our associating ideas via *probable* reasoning.

With that, I hope to have tied up several (but not all) of the loose ends raised in the previous two chapters. The faculty of reason comprises not only demonstrative reasoning and induction, but also a recombinatory faculty that functions to explain the inductively-established universal regularities that we discover in experience. Because it functions to explain, and is thus responsible to experience not only for its representational content, but also as the standard for its explanatory success, reason, unlike the imagination, is a reliable recombinatory faculty, whose representations are relatively stable, and thus accrue a high degree of force and vivacity. These representations, in turn, represent the nature, power, and essence of the substance that underlies manifest phenomena. As such, they represent these manifest phenomena as being the result of the conglomerate distinctly *causal powers* of this underlying substance.[30]

That is the big picture of how I understand Hume's scientific realism. I have already considered at least three successful applications of this picture. The impression-idea distinction, the simple-complex distinction, and the distinctions between memory, imagination, and reason all represent real distinction among our perceptions and faculties the representation of which are the result of reason's recombining the data of experience (perceptible models) into representations with novel content that function to explain some universal regularities of experience. I have argued that Hume does, and must, take these distinctions to represent the real nature of the human mind. This picture, however, not only predicts the form that explanatory hypotheses and theoretical posits will take when successful, but also the ways in which these can fail. In particular, when such hypotheses are undertaken not on the basis of reason, but as a result of the imagination's influence, they will have neither the epistemic *bona fides* nor the resultant truth-tracking reliability to prove tenable. Or, when such posits are not formed by specifying the determinate ways that the theoretical entity at hand is meant to resemble and differ from its perceptible model, they can have neither the representational content necessary to be genuinely explanatory, nor the explanatory force that comes from subsuming a conceiving one phenomena as of a kind with some antecedently understood one. It is to the examination of such failed cases that I will turn in the final three chapters: the attempts to explain the apparent constancy and coherence of our perceptions, the attempt to represent necessary connection, and the attempt to explain the unity of consciousness. In each of these cases, an examination of the failures of the attempt at explanation at hand will both confirm the picture I have been giving of Hume's scientific realism, but also reveal important nuances of that view, and of course, yield surprising conclusions about the extent of our representational powers and understanding of the nature of the human mind.

Notes

1. Summarizing Ayers on naturalism, Kail writes, "All agree that God is the 'first cause' in that He created the world. The issue was whether He also created 'second causes', or objects with natures in virtue of which they bring about their effects, or whether he set up laws of nature among powerless natural objects, which are maintained by God's efficacious will. Naturalism has it that there are second causes. Natural objects have causal powers in a way that relates to their status as natural kinds" (Kail, *Projection and Realism*, 85–6). Thus, the view that I will defend here can be counted as naturalist in this respect.
2. As we will see in the final section of this chapter, Hume uses "imagination" in a broad and narrow sense. Construed broadly, it is the name of any faculty for recombining simple ideas to form representations with novel content. In that sense, it contrasts with memory, which is reproductive rather than recombinatory, and includes reason as a sub-species. Construed narrowly, it is the name for a recombinatory faculty the operation of which is unconstrained by experience,

or not answerable to the truth. In that sense, it contrasts most directly with reason. In this passage, Hume is using "imagination" in the former, broad sense.

3. Notice that the role of science here is to discover which qualities are part of the nature of gold!
4. Hume's endorsement of inference to best explanation here is a striking contrast to Bas van Fraassen's rejection of the realist consequences of that same practice. van Fraassen, *The Scientific Image*, 20–1.
5. Jonathan Cottrell has raised the following objection. At Abs. 28 Hume writes,

> [I]t must be our several particular perceptions that compose the mind. I say *compose* the mind, not *belong* to it. The mind is not a substance in which the perceptions inhere.
>
> Abs. 28

This text appears to contradict my thesis that Hume takes science of human nature to explain the regularities of experience via an appeal to the substance of the mind rather directly. Hume writes that *the mind is not a substance*. Notice first, however, that what Hume actually claims here is that the mind is not a substance *in which perceptions inhere*. That is true on my interpretation of Hume because, as he writes here, the substance of the mind is *composed* of perceptions. As I argued in Chapter 1, simple perceptions are themselves theoretical posits meant to explain the variety of human thought in the fact of the relative paucity of human experience. As such, they themselves collectively compose the substance of the human mind, and so cannot also inhere in it.

Interestingly, the remainder of this paragraph from the Abstract does contain what appears to be a more wholesale rejection of the notion of substance, but which actually follows the same pattern that we have just observed in 1.1.6. Hume rejects only the notion of substance *as a via negativa*, but also retains his own conception of the proper representation of substance as being modeled perceptions themselves.

> That notion is as unintelligible as the *Cartesian*, that thought or perception in general is the essence of the mind. We have no idea of substance of any kind, since we have no idea but what is derived from some impression, and we have no impression of any substance either material or spiritual. We know nothing but particular qualities and perceptions. As our idea of any body, a peach, for instance, is only that of a particular taste, colour, figure, size, consistence, *&c*. So our idea of any mind is only that of particular perceptions, without the notion of any thing we call substance, either simple or compound.
>
> Abs. 28

While Hume does write here that, "we have no idea of substance of any kind," since we have just seen him give *his own account* of the idea of substance, what he must mean by that is only that we have no idea of substance *as incorrectly conceived by his predecessors* (e.g., Cartesians). The scope of the phrase "of any kind" is not meant to encompass any conception of substance (including his own), but rather only the material and spiritual substances directly referenced at the end of that sentence.

6. Sellars, *Science and Metaphysics*, §90.
7. It is worth noting that the language of the science of human nature, or the true philosophy, has a competitor to replace the language of the vulgar: that of the false philosophy. I postpone discussion of the false philosophy to the next chapter.

8. This way of putting the issue is admittedly an over-simplication. While the vulgar's appeal to body might not provide a uniting principle as the foundation for the general term "dog", *biologists*, for example, might take certain facts about dogs' phylogenetic place to do so. If biology is a successful science, then, "dog," as used in the language of biology, will represent a substance. Of course, since it is the science of human nature that is most explanatorily fundamental, it is also that science that serves as the ultimate arbiter of questions of ontological commitment. Whether the science of human nature ultimately validates or vitiates the ontology of the other sciences is a question for a future study.
9. Hume also refers to this as men's "common and careless way of thinking" (1.4.3.9; SBN 223).
10. Butler disagrees.

    > Although Hume introduces the general idea of identity, we should not conclude that thinking of some particular thing as one and the same thing over time requires general ideas. Instead, Hume aims to describe in general the complex structure that any idea of a particular identical object will exhibit.
    >
    > Butler, "Hume on Believing the Vulgar Fiction of Continued Existence," 239.

    My argument to the contrary is that Hume is explicit that identity across time requires supposing the qualities of an object to have a common cause in virtue of its being the *kind* of object that it is (T 1.3.2.2; SBN 74), and that his account of substance implies both that a substance is a set of qualities with a common cause and that objects are represented as certain kinds of substance via general terms. Thus, representing identity across time does require the use of general terms.
11. Similarly, Hume notes in passing in 1.3.14 that the common distinction between moral and physical necessity is "without any foundation in nature," i.e., does not reflect the most explanatory ontology.

    > The same course of reasoning will make us conclude, that there is but one kind of *necessity*, as there is but one kind of cause, and that the common distinction betwixt *moral* and *physical* necessity is without any foundation in nature.
    >
    > T 1.3.14.33; SBN 171

    The best explanation of the relation of cause and effect posits just one kind of cause, not two, and so is the former ontology is what reflects nature.
12. Flage, "Hume on Memory and Causation," argues that memory necessarily reproduces the content and structure of impressions. Traiger, "Flage on Hume's Account of Memory," objects that since simple impressions and ideas have no structure, Flage's thesis would imply that simple ideas cannot be memories and that simple impressions cannot be remembered. Flage, "Perchance to Dream," rightly notices that since we never have impressions or ideas of simple ideas *in isolation*, Traiger's objection gains no purchase. One might also argue, with texts to support it (e.g., T n5; SBN 637), that their simplicity itself is a kind of structure that is reproduced by memory. In the later paper Flage also there answers three other of Traiger's objections.
13. It is important to note that the notion of reproduction at work here includes both that memories replicate the order and position of our impressions, and that they are *caused by* those impressions. That is, an imagining that just so happens to correspond to some past impression is not thereby a memory of it. The complement of this fact is that inaccurate memories will be caused by some past impression, but will fail to resemble it. This raises the thorny issue

of how Hume can demarcate the difference between inaccurate memories and imaginings. My answer to that question is that we ought to take Hume's talk of these representations being produced by different faculties seriously. It is the *function* of memory to reproduce; imagination has so such function. Construing this difference in this way allows for distinguishing inaccurate memories from imaginings by appeal to the underlying faculty that produces them.

By contrast, David Owen disavows Hume's idiom of faculty psychology as, "against the whole spirit of his theory" (Owen, *Hume's Reason*, 76). As Owen sees it, that spirit is embodied by Hume's supposed "refusal to go beyond experience" (Owen, *Hume's Reason*, 6, 9, 65, 76). Perhaps because this latter thesis has been a mainstay of Hume scholarship for some time, Owen presents little textual evidence in support of it. (He does cite two passages—Intro. 8 and 9—that I argued in Chapter 2 do not support this conclusion.) One upshot of the view of reason for which I will argue in the next section is that it allows us to understand Hume as having a more sophisticated understanding of the role of experience in epistemic justification than does this standard reading. Regarding reason specifically, Owen does offer the following argument.

> If the faculties of memory and the imagination are really distinct, and the understanding is a sub-faculty of the latter, then the understanding should deal with nothing but ideas of the imagination. But, of course, our reasoning makes use not only of ideas of memory as well as those of the imagination, but also of impressions (as in the 'inference from the impression to the idea').
>
> Owen, *Hume's Reason*, 76.

It is easy enough, I think, to reject the first premise of this argument: the conditional that if reason is a sub-faculty of the imagination, then it should deal with nothing but imaginings. As I understand them, faculties are defined by their proper functions, which are the operations that they normally perform on perceptions, not by which perceptions they operate on. So, for example, I understand memory as functioning to reproduce. Paradigmatically, that is the reproduction of the structure and content of impressions, but can of course also reproduce imaginings, such as when one remembers a dream. By contrast, the imagination is a recombinatory faculty that operates under no constraints, and it deals with impressions, memories, previous imaginings, etc. I fail to see a contradiction here. (Of course what a faculty *produces* will, of necessity, be products of that faculty—memory produces memories, imagination imaginings, etc.—but that does not appear to be what Owen means).

14. "Weakly" because he never comes out and says that this is a *criterion* of the difference, yet does cite it as an evident universal feature of memories, and perhaps as a necessary condition of being a memory.
15. That Hume is interested in distinguishing all three of these kinds of representations is first emphasized in Friedman, "Another Look at Flage's Hume."
16. McDonough, "Hume's Account of Memory," 80.
17. McDonough, "Hume's Account of Memory," 82–3.
18. In fact, Hume is explicit not only that the resemblance between an impression and an idea transfers some of the force and vivacity of the former to the latter, but also that a similar transfer occurs even in cases where such a resemblance is merely *intended*.

> I wou'd willingly establish it as a general maxim in the science of human nature, that when any impression becomes present to us, it not only

> transports the mind to such ideas as are related to it, but likewise communicates to them a share of its force and vivacity. [. . .] We may, therefore, observe, as the first experiment to our present purpose, that upon the appearance of the picture of an absent friend, our idea of him is evidently enliven'd by the resemblance [. . .] Where the picture bears him no resemblance, or at least was not intended for him, it never so much as conveys our thought to him
>
> T 1.3.8.2–3; SBN 98–9.

Even if a picture does not resemble our friend at all, if it is intended to resemble him, our thought can still be conveyed to him. This remark suggests that Hume is comfortable, at least to some degree, with describing our mental faculties in functional terms. ("Intention" and its cognates appear more than a dozen times in the *Treatise*, never with any hint of skepticism.) That is important for current purposes insofar as it is my thesis that Hume understands memory as *functioning* to reproduce (and will later be that reason as *functions to* explain). Of course, it may not be obvious how to give cash value to this functional idiom in strictly Humean terms, even if Hume himself endorses it, but a good deal of contemporary literature, for example in the philosophy of biology, attempts to do so. The *locus classicus* for a naturalistic account of biological functions is Millikan, *Language, Thought and Other Biological Categories*.

19. For example, see T 1.3.9.6; SBN 109–10.
20. Of course, because our cognitive faculties are defined functionally, it will sometimes occur that while memory's *function* is to reproduce the structure of our impressions—that is it's "chief exercise"—it will sometimes fail in that task: we misremember. There are a variety of ways that this can happen, several of which Hume comments on, e.g., the influence of education or habit on our thinking (T 1.3.9.16–19; SBN 115–17), or the tendency of the mind, "when set into any train of thinking, [. . .] to continue, even when its object fails it" (T 1.4.2.22; SBN 198).
21. Garrett, *Cognition and Commitment*, 21.
22. Garrett, *Cognition and Commitment*, 21.
23. In fairness, Garrett does elsewhere distinguish between the imagination broadly construed, which includes reason as a sub-faculty, and the imagination narrowly construed, which is also a sub-faculty of the imagination broadly construed and is distinct from reason. The question, then, is what constitutes the distinctions between memory and imagination broadly construed, on the one hand, and reason and imagination narrowly construed on the other. I will attempt to answer those questions in a moment, and I will argue that it is most accurate to construe Hume as countenancing *three* faculties here (defined by their respective functions).
24. Loeb does an excellent job of making the case that whereas Hume sometimes uses "reason" or "the understanding" to describe a non-associative faculty the existence of which he rejects, he also appropriates that term to describe his own conception of reason as a valid principle of the imagination (broadly construed). Loeb, *Stability and Justification*, 53–9. Thus, as I will argue in a moment, Loeb gets right that reason is a reliable recombinatory faculty, but misses the important point that its main function is to *explain*.
25. Of course, explaining empirically discovered regularities of perception cannot be the *only* function that reason performs. For example, *inductive* reasoning is necessary to first reveal such regularities, and *demonstrable* reasoning, while explanatory in some contexts, is explicitly excluded from certain aspects of causal explanation by Hume. Thus, to put it more carefully, my thesis is

that there is a *kind* of reasoning the function of which is to explain inductive-established universal regularities of experience. This kind of reasoning is what has been called inference to the best explanation, or abductive inference. Hume sometimes appears to include it under the rubric "probable reasoning", but it is clearly distinct from more straightforward inductive inferences.

26. While I take the benefit of the stability of belief to be its indicating *truth*, which I take to be the ultimate aim of the science of human nature, there is still much to be said for Loeb's excellent account of epistemic justification as, "deriving from the motivational force of the felt uneasiness in unstable doxastic conditions" (Loeb, *Stability and Justification*, 22). While I disagree with Loeb that the benefit of stability derives primarily from its elimination of the feeling of uneasiness, I agree with him on the importance of stability in distinguishing reason from imagination, custom, habit, and education, and in Hume's systematic account of belief and belief-forming mechanisms. It should be noted that Loeb does outline a sense in which belief aims at truth (Loeb, *Stability and Justification*, 82–7), but this sense is still secondary to belief's primary aim of stability.
27. In fact, Hume's inference here implies the stronger claim that reason cannot err at all. That strong thesis, though, cannot be one that Hume actually holds. (Cf. 1.4.1 and 1.4.2.) Therefore, I take this passage to comprise some hyperbole, the underlying truth of which is that reason reliably tracks the truth whereas imagination does not.
28. Garrett, *Cognition and Commitment*, 92.
29. Owen reads Hume this way, emphasizing that he differs from Locke insofar as Hume understands probable inference as a matter of moving from one idea to the next without the need for an intermediate idea that is intuited as explaining the transition. Owen, *Hume's Reason*, 125–33.
30. One final objection: if Hume endorses a form of inference to the best explanation, then what are we to make of his arguments that appear to proceed by ruling out demonstrable and probable reasoning, and then concluding that reason is not responsible for some phenomenon or other? Such arguments suddenly appear enthymematic, missing a crucial premise that rules out inference to the best explanation as well. Again, the devil is in the details of each of these arguments, and I will address some but not all of them in what follows. (For example, in Chapter 6 I show that Hume's argument that we have no idea of the self explicitly takes into account the possibility of forming a theoretical representation of it.) Generally speaking, there are a few interpretive principles that can address these arguments. First, sometimes Hume understands "probable reasoning" broadly enough to include inference to the best explanation, for example in 2.3.3 where he argues that reason cannot motivate. Second, sometimes it is true that Hume does seem to neglect this as an option, but it is clear how the argument would go, had he considered it. Finally, it is worth noting that whatever pressure such examples put on the attribution of an endorsement of inference to the best explanation to Hume, are counterbalanced by his apparent *use* of that form of explanation.

4 The Science of Body

Having explicated Hume's scientific realism, and seen several important examples of its positive application in the *Treatise*, I can now articulate in broad outline Hume's methodology in conducting the science of human nature, at least in those cases where that methodology leads to the successful explanation of some phenomena.

- The need for an explanation arises from the discovery, by reason, that observed particulars obey certain universal empirical regularities.
- The needed explanation of these regularities casts them as the result of the nature, powers, essence, or *substance* that the observed particulars really are.
- The idea of such a substance is formed by specifying the determinate ways that some observed phenomena (perceptible model) both resemble and differ from a posited theoretical-explanatory entity.

In the first chapter, we saw this procedure play out in two important cases: the impression-idea distinction and the simple-complex distinction. In the third we saw it at work in the forming the distinctions between imagination, memory, and reason. What we will see in this chapter is that the pursuit of the science of human nature does not always proceed so smoothly. In particular, what we find in the arguments spanning 1.2.6, 1.4.2, 1.4.3, and 1.4.4 is that while something very much like this procedure can be found in both the vulgar's and false philosophers' attempts to explain the apparent constancy and coherence of our perceptions by making use of a perceptible model—perceptions themselves—both of these attempts fail in precisely the ways that this understanding of scientific methodology predicts. Specifically:

- Neither explanation is the result of *reasoning* about the phenomena at hand, but is rather a product of the imagination.
- Neither attempt at explanation is capable of specifying *determinate* differences and similarities between the model observable phenomena and the theoretical entity posited.

- The explanation of the vulgar straightforwardly contradicts empirical evidence.
- The explanation of the false philosophy, while it ostensibly replaces the language of the vulgar, is really entirely parasitic on it, i.e., only replaces the *words*, but not the ideas or associations that give these words their meaning.

Of course, with these failures of explanation the *explicandum* remains: why *is* it that our perceptions appear to exhibit such steadfast constancy and coherency? What Hume discovers in the course of his investigations is that these apparent regularities are *merely* apparent: they are found not by reason, but are "discovered" by the imagination, which has an unfortunate tendency to fall into habits that cause it to exaggerate the small and mundane regularities that in fact exist. Thus, the *explicandum* of the systems of the vulgar and the false philosophy are illusory, and the search for its *explanans* ought to be abandoned. That, however, proves to be more difficult than it would seem. As Hume notes at the start of 1.4.2,

> We may well ask, *What causes induce us to believe in the existence of body?* but 'tis in vain to ask, *Whether there be body or not?* That is a point, which we must take for granted in all our reasonings.
>
> T 1.4.2.1; SBN 187

Despite the failure of any attempt to explain the constancy and coherence of our perceptions, and even with the realization that there is nothing there to explain, we find ourselves compelled to believe that there is *something* that can do this explanatory work. It is the resilience of that belief, in turn, that is itself discovered to be a universal empirical regularity that calls for its own explanation, which is what Hume undertakes to give.

What I argue here is that the interpretation that I have been defending predicts not only Hume's own important successes in pursuing the science of human nature, but also his account of the failures of his predecessors. The *explicandum* of both the vulgar's and false philosophy's systems are not the particulars themselves (perceptions), but rather the fact that these particulars obey the universal empirical generalizations that they do. The purported *explanans* of these empirical generalizations is not some further, higher-order generalization, but an appeal to the nature, power, and essence of the substance underlying them. In the case of the vulgar, the posited substance is a perception that is distinct from the mind and continues to exist when not perceived. In the case of the false philosophy, it is a material substratum that in no way resembles the perceptions that represent it (has no perceptible model). And, of course, Hume does not reject the realist implications of these attempts at explanation *per se*, but only their specific executions.

My procedure will be as follows. In the first section I will review the arguments from 1.2.6 that Hume reports frame the discussion of body in

1.4.2. What I will show is that these arguments are intended to demonstrate not only that we find ourselves with no idea of external existence, but also that there is no possible theoretical posit that can do the work of representing such a thing. Such a posit would not resemble perceptions in any way at all, i.e., would not be founded or *modeled* on anything in experience, and would thus amount to a mere *via negativa*. I have been defending the view that Hume allows for no such representations, but there are Hume scholars, most notably the proponents of the New Hume, that disagree, i.e., who maintain that Hume holds that we can refer to certain objects even without being able to form an idea of them with any descriptive content. In the second section of this chapter, I turn to one prominent defense of that position, Galen Strawson's, to show where I take it to go wrong. With that framework established, the third section explicates the failures of the systems of the vulgar and false philosophy. I there argue that the understanding of Hume's pursuit of the science of human nature that I have been defending best explains both what the vulgar and false philosophy *take themselves* to be doing, and why they fail. In the final section, I do the same for Hume's own explanation, not of the supposed constancy and coherence of our perceptions, but of our persistent belief in these regularities.

The Ideas of Existence and External Existence and the Copy Principle

Before we proceed to our examination of the first four sections of 1.4, "Of the sceptical and other systems of philosophy," it will be worthwhile to first consider the important thesis that Hume himself cites as circumscribing that discussion:

> For as to the notion of external existence, when taken for something specifically different from our perceptions, we have already shown its absurdity.
>
> T 1.4.2.2; SBN 188

Certainly many philosophers both before and after Hume thought it appropriate to begin their discussion of what Hume calls "body" by contrasting body with the mental. As is well known, Hume rejects such a contrast, instead casting the vulgar idea of body as the idea of a *perception's* being distinct from a mind and continuing to exist when not perceived. We will return to this understanding in a moment, but we must first understand the grounds on which he rejects the notion of something "specifically different from our perceptions" as absurd. That claim refers back to the arguments of 1.2.6, "Of the idea of existence, and of external existence," and there are several striking features of the argument that one there finds. First, the section is very short—nine paragraphs in all—which is less than one would expect it would take to demonstrate that all that we can ever think about is our own perceptions. Second, even that small space is divided between *two* arguments, the first of which is for the conclusion that "[t]he idea of existence [. . .] is the very same with the idea of what we conceive to be existent." Third, after completing the argument regarding the idea of existence, Hume notes that, "[a] like reasoning will account for the idea of *external existence*," but it is not at all obvious what the single argument form it is that is supposed to be operative in both arguments. Finally, it is not *prima facie* clear that either of the arguments of this section is even valid, much less sound.

So, before we can even so much as approach the discussion of the explanations of the vulgar and of the false philosophy, we have before us the task of making sense of this very short, dense, and puzzling section of the *Treatise* in which Hume argues for at least one incredibly audacious conclusion. I will take as my entry point into this task Hume's own clue that the two arguments of this section share a common argumentative form. On a first reading, that common form appears to involve an essential appeal to the Copy Principle, but what I will show is that the Copy Principle alone is not enough to make these arguments good. Hume's arguments also require a premise that validates the move from a claim about the causal origin of an idea, to a claim about that idea's representational content. Once the need for such a principle is introduced, however, so is the possibility of representing existence or external existence via a *theoretical* representation. With that in mind, a careful reconsideration of the text reveals that it is precisely the thesis that I have been attributing to Hume—that theoretical posits are represented via perceptible models that they resemble and differ from in determinate ways—to which Hume appeals to demonstrate that there can be no such representation of either existence of external existence.

We can begin, then, with Hume's argument for the conclusion that the idea of existence is the very same idea as the idea of what we conceive to be existent. That argument begins with the premise that all of our ideas are ideas of their objects as existing.

> There is no impression nor idea of any kind, of which we have any consciousness or memory, that is not conceiv'd as existent;
>
> 1.2.6.2; SBN 66

Next, Hume claims that it is from such ideas that we derive our idea of "being."

> and 'tis evident, that from this consciousness the most perfect idea and assurance of *being* is deriv'd.
>
> 1.2.6.2; SBN 66

Hume next invokes to the Copy Principle to derive a dilemma from this claim.[1] Since every simple idea is copied from some impression, and the idea of being (or existence) can be derived from any perception, either there is some impression of being that accompanies all of our other perceptions, or the idea of being is identical to those perceptions.

> From hence we may form a dilemma, the most clear and conclusive that can be imagin'd, *viz.* that since we never remember any idea or impression without attributing existence to it, the idea of existence must either be deriv'd from a distinct impression, conjoin'd with every perception or object of our thought, or must be the very same with the idea of the perception or object.
>
> 1.2.6.2; SBN 66

Next, Hume invokes the Separability Principle to rule out the first horn of this dilemma.

> So far from there being any distinct impression, attending every impression and every idea, that I do not think there are any two distinct impressions, which are inseparably conjoin'd. Tho' certain sensations may at one time be united, we quickly find they admit of a separation, and may be presented apart. And thus, tho' every impression and idea we remember be consider'd as existent, the idea of existence is not deriv'd from any particular impression.
>
> 1.2.6.3; SBN 66

Because any two distinct impressions are separable, if there were a distinct impression of existence, it would be separable from any of the perceptions that it is supposed to accompany. Finding that our idea of existence is in fact *not* separable from the idea of some object,[2] Hume concludes that the second horn of the dilemma must be true: that our idea of existence is identical to the idea of the object conceived to exist.

> The idea of existence, then, is the very same with the idea of what we conceive to be existent. To reflect on any thing simply, and to reflect on it as existent, are nothing different from each other. That idea, when conjoin'd with the idea of any object, makes no addition to it. Whatever we conceive, we conceive to be existent. Any idea we please to form is the idea of a being; and the idea of a being is any idea we please to form.
>
> T 1.2.6.4; SBN 66–7

While the details of this argument are far from clear or conclusive, this outline of it should be enough to shed light on the form of Hume's second argument from this section: that concerning external existence. (Recall that Hume claims that, "A like reasoning will account for the idea of *external existence*" (T 1.2.6.7; SBN 67).)

So, what is the form of the argument that we have just considered? Roughly and in broad strokes, it looks something like the following.

(1) Every idea is an idea of something existent.
(2) Our idea of existence can be derived from any idea. (1)
(3) The Copy Principle.
(4) There must be some impression of which our idea of existence is a copy. (3)
(5) Either our idea of existence is a copy of an impression that accompanies all other perceptions, or it is a copy of those perceptions themselves. (2, 4)
(6) Our idea of existence is not a copy of an impression that accompanies all other perceptions. (via the Separability Principle)
(7) Our idea of existence is a copy of perceptions themselves. (5, 6)
(8) "The idea of existence, then, is the very same with the idea of what we conceive to be existent." (7)

With this rough reconstruction as a guide, we can now turn to the argument concerning external existence, to see how it is supposed to function. Again, roughly and broadly, here is that argument.

(1) "We may observe, that 'tis universally allow'd by philosophers, and is besides pretty obvious of itself, that nothing is ever really present with the mind but its perceptions or impressions and ideas, and that external objects become known to us only by those perceptions they occasion. To hate, to love, to think, to feel, to see; all this is nothing but to perceive." T 1.2.6.7; SBN 67
(2) "Now since nothing is ever present to the mind but perceptions"
(3) "and since all ideas are deriv'd from something antecedently present to the mind"
(4) "it follows, that 'tis impossible for us so much as to conceive or form an idea of any thing specifically different from ideas and impressions." T 1.2.6.8; SBN 67 (1, 2, 3)

Hmm. That argument is much shorter and more straightforward than its predecessor. It does not proceed via a dilemma, does not make use of the Separability Principle, and it appears to deploy the Copy Principle in a much more straightforward way. So, in what sense does "a like reasoning [. . .] account for the idea of external existence"? Well, the common element appears to be that in both arguments the Copy Principle is used to show that each of the respective ideas of existence can only be ideas of certain perceptions themselves. More specifically, in both cases it facilitates the inference from the claim that we have no existence-impression or external-existence-impression itself to the relevant conclusion about our idea of such existences. Hume takes a little more time and care in the first argument establishing the crucial basis for this inference, but perhaps that is simply because having done so in that first argument, he does not need to repeat the process again in the second one.

So the similarity between the two arguments comes down to the application of the Copy Principle. As I argued in Chapter 1, though, it is not *just* the Copy Principle on which Hume relies in such arguments, but also what I have called the *Representational* Copy Principle. We previously encountered that thesis in the context of understanding Hume's theory of general representation, but it is worth revisiting here given its importance in interpreting the arguments at hand.

Recall that we noted the close tie between Hume's Copy Principle and the theory of representation that appears to make use of that principle. The Copy Principle states that all simple ideas are caused by and exactly resemble some simple impression. The corresponding theory of representation—the Representational Copy Principle—states that all simple ideas *represent* the simple impressions of which they are copies. Noting that this latter principle would not be sufficient for accounting for either misrepresentation or complex representation, we amended it to include these phenomena by adding that a complex representation represents the objects of its simple component representations as being arranged as those simple representations are arranged in that complex. Schematically,

> 'x'R'y' represents xRy.

A representation of x and y as related to one another in way R consists of a representation of x related in the same way to a representation of y. We noted at the time that since these component simple ideas will be copies of the simple impression that they represent, when a complex idea *correctly* represents an arrangement of simple impressions, it too will be a copy of those impressions. That is, a complex idea is correct just in case it is caused by and exactly resembles the arrangement of simple impressions that it represents. It will be important in the current context, though, to carefully distinguish the Copy Principle from its representational counterpart. Consider again the following picture.

What I have been calling the *pictorial character* of this picture consists of four black lines of equal length arranged at ninety-degree angles to one another against a white background. That is in what *the picture* consists. For another picture to exactly resemble this one, it would also have to consist in four lines of this length arranged at ninety-degree angles to one another against a white background. The pictorial character of an image, including impressions and ideas, is constituted entirely by the *intrinsic* features of that image. In describing the image above, though, we made no reference to what that image is an image *of*. That is, we described the intrinsic features of that picture, but did not mention, for instance, that it is a picture of a building as seen from directly above, or a book seen

Figure 4.1

from straight on, etc. What a picture is a picture *of* is the *representational content* of that image, and is pre-theoretically possible for the pictorial character of an image and the representational content of that same image to come apart.

So, Hume's Copy Principle states that all ideas are copies of impressions, i.e., that all ideas are caused by, and have exactly the same *pictorial character* as, some corresponding impression. Important for us to note is that, as formulated here, the Copy Principle does not speak at all towards the *representational* content of impressions or ideas. It is merely a thesis concerning the causal relations between impressions and ideas and the relation of their pictorial character. That fact is important in the current context because the conclusions that Hume reaches in the arguments of 1.2.6 are explicitly concerned with the representational content of the ideas of existence and external existence.

> The idea *of* existence, then, is the very same with the idea *of* what we conceive to be existent. To reflect on any thing simply, and to reflect on it *as* existent, are nothing different from each other. [. . .] Any idea we please to form is the idea *of* a being, and the idea *of* a being is any idea we please to form.
>
> T 1.2.6.4; SBN 66–7, emphasis mine

> 'tis impossible for us so much as to *conceive* of form any idea *of* any thing specifically different from ideas and impressions.
>
> T 1.2.6.8; SBN 67, emphasis mine

Since it is at conclusions about the representational content of our ideas at which Hume's arguments aim, despite their surface structure, these arguments must implicitly be employing some premise or premises concerning not just pictorial character, but also representational content. Since Hume's use of the Copy Principle is explicit, what is needed to make these arguments good is a thesis that links the pictorial character of an idea to its representational content. The most plausible such thesis is what I have called the Representational Copy Principle, which states that the representational content of any perception is that of which it is a copy. To put it another way, the Representational Copy Principle states that a perception is *of* that of which it is a copy. It is *this* thesis that allows us to move from the claim that we do not have an idea with a certain pictorial character to the claim that we do not have an idea with certain, corresponding representational content. In particular, it would allow the move from, for instance,

(1) "We have no idea that is a copy of an external existence"

to

(2) "We have no idea that is *of* an external existence".

(2) is what Hume is after in 1.2.6. (1) is merely a necessary step along the way. (1) is what the Copy Principle earns Hume. (2) is only earned via the Representational Copy Principle.

Thus, Hume's use of the Copy Principle in the arguments concerning existence and external existence is more complicated than it first seems in at least two ways. First, it is not just the Copy Principle at work in those arguments, but also the Representational Copy Principle. If Hume is going to be justified in concluding that we have no distinct idea *of* existence or *of* external existence, then he will need to rely on a thesis about how it is that our ideas come to have the representational content that they do. Second, even such a principle does not validate the *direct* move from, e.g.,

(1) There is no simple impression from which our idea of external existence is a copy.

to

(2) There is no idea of external existence.

because any tenable thesis about the representational content of our ideas will of necessity include the possibility of complex representational content, and so will leave open the possibility that even though our idea of, e.g., external existence is not a copy of any simple impression, it could still be an arrangement of simple ideas that are themselves copies of some simple impressions that we do have.

Finally, there is a third possibility for which Hume's argument must account: the possibility of using a perceptible model to represent a theoretical posit, which is itself never experienced. As we noted in the first chapter, our general idea of simple ideas has as the members of its revival set certain, but not other, complex ideas. The complex idea of watching a spot of ink on a piece of paper disappear as one walks away from it is in. The intricate mechanisms one discovers upon opening the back of one's watch is out, etc. One question concerning such revival sets is how they come to be. How do we come to associate the former but not latter with the term "simple idea"? What makes it so that the revival set of the term "simple idea" contains these ideas but not others? Here is where the requirement that a theoretical posit have a perceptible model that it resembles and differs from in determinately specifiable ways plays a crucial role. As we saw, Hume specifies the content of the revival set of "simple idea" by taking complex ideas as his model and specifying that simple ideas are like these in some ways (e.g., they are images), and unlike them in others (e.g., they cannot be distinguished into parts). A perceptible model is thus the key to fixing the content of the revival set of any new general term. Since the object of a theoretical term cannot be perceived, the only way to represent it is by contrasting it with that which can be perceived, and what is posited by such a contrast is a *real difference* between the members of this contrastive revival set and all others. That is, in adopting the term "simple idea" one thereby endorses the thesis that the members of the revival set of "simple idea" differ in some real way from other ideas (complex ones). It is through the use of a perceptible model, by contrasting the revival set of "simple idea" with that of a general term that includes perceptible objects as its members ('complex idea') in *specific* ways that "simple idea" comes to have the *determinate* content that it does. (This will be a particularly important point to keep in mind in the next section, on so-called "relative ideas.")

So, the final possibility for which Hume must account in arguing against the possibility of the ideas of existence and external existence is that we could form an idea of either of these via the deployment of a theoretical posit. This last notion is perhaps the most important one to be on the lookout for in Hume's arguments here, because while the issue of representational content *per se* is crucial, it is also one that Hume often leaves implicit. That is, many of Hume's conclusions throughout the *Treatise* are explicitly representational, but the arguments for most of these conclusions leave implicit the step wherein the argument moves from the application of the Copy Principle to the application of the Representational Copy Principle. The second concern here, about the possibility of a there being a complex idea of existence or external existence constructed from more mundane simple ideas can also be mollified. The worry here is that it is not sufficient to show that we do not have a certain idea that one shows that that idea could not be copied from some impressions because most of our complex ideas are not copied from any impression, and yet there they are. The obvious reply to this concern is that if Hume is successful in showing that there is no *simple* impression of existence or external existence, he also thereby shows that no *combination* of ideas of the simple impressions that we do have will have as part of its content any distinct impression of existence or external existence. This response would work perfectly well, if it weren't for the possibility of complex theoretical representation that we have recently noted, and so once again that is the possibility on which both arguments will turn.

So, if these arguments are to work, Hume will have to rule out the possibility of using a general term to represent a distinct idea of existence or of external existence. And lo and behold, that is precisely what Hume does in the paragraph between the end of the argument against the former and beginning of the argument against the latter.

> Our foregoing reasoning concerning the *distinction* of ideas without any real *difference* will not here serve us in any stead. That kind of distinction is founded on the different resemblances, which the same simple idea may have to several different ideas. But no object can be presented resembling some object with respect to its existence, and different from others in the same particular; since every object, that is presented, must necessarily be existent.
>
> T 1.2.6.6; SBN 67

Hume is here referencing his discussion of distinctions of reason in 1.1.7, which we observed previously is a straightforward application of the Representational Copy Principle as it applies to resemblance relations, a thesis that Hume himself here endorses: "that kind of distinction is founded on the different resemblances, which the same simple idea may have to several different ideas." As I have emphasized, using a general term to form a representation of a theoretical entity requires specifying the similarities and the differences from some perceptible model. In most cases that Hume examines, this requirement is violated by philosophers that attempt to represent that which resembles our perceptions *in no way*, i.e., by philosophers who posit a mere *via negativa* as explaining some manifest phenomena. As we can see from the above quotation, however, there is complementary way to violate this principle: by positing that which *differs* from our perceptions in no way. That is, Hume here completes his argument against the distinct idea of existence by pointing out that the representational content of the theoretical posit suggested by the proponents of such an idea would be *identical* to the representational content of its intended *explicandum*. Since every idea is the idea of its object as existing, there is no way to specify any difference between an idea of an object and an idea of the existence of that object. Whatever the ultimate worth of this argument, it is striking that Hume presents it at exactly the moment when what he needs to complete his argument against the distinct idea of existence is a way of ruling out the possibility that this idea, like the distinctions between "simple idea" and "complex idea" and "impression" and "idea", represents certain but not other of our ideas as bearing real resemblance relations to each other. Unlike his own distinctions, there is no room to use a perceptible model to represent existence precisely because there is no way to specify any difference between such a model and any possible theoretical posit.

The argument against the distinct idea of external existence leverages the more common point that Hume makes about perceptible models. What Hume objects to in that case is the suggestion that we form an idea of external existence that is specifically different from our perceptions. That is, Hume's objection is that we cannot form an idea that in no way resembles our perceptions. We *can* form ideas that differ from our perceptions in determinate and specifiable ways. The idea of external existence, however, would be one that in no way resembled any perception.[3] Again, it would be a mere *via negativa*, which we have seen for Hume is no representation at all. Thus, he concludes that the most that we can do by way explicating the notion of something entirely different from our perceptions is to make an incomprehensible gesture at them.

> The farthest we can go towards a conception of external objects, when suppos'd *specifically* different from our perceptions, is to form a relative idea of them, without pretending to comprehend the related objects.
>
> T 1.2.6.9; SBN 68

We will consider this passage in more detail in the next section, but what I want to emphasize for the moment is that we are now in a position to answer the question of in what way the two arguments of this section—against the distinct ideas of existence and external existence—share a form. Both conclusions explicitly concern the *representational* content of our ideas. Both arguments must therefore make use of a thesis concerning how our ideas come to have the representational content that they do. That thesis is the Representational Copy Principle, the correct deployment of which requires that Hume not only use the Copy Principle to show that we have no *simple* impression of which such ideas could be a copy, but also that no *arrangement* of the simple ideas that we do have is up to this task, and also that no *theoretical* representation is either. In the end, Hume does precisely this, by showing—as we have previously observed him do in many other cases—that such ideas would violate the single most important principle of theoretical representation: that all such representation employ a perceptible model, the determinate differences from and similarities to which are what constitute the representation of the theoretical posit at hand.

The candidate ideas here violate that principle in complementary ways: the distinct idea of existence cannot *differ* from any perceptible model in determinate ways, whereas the idea of external existence cannot *resemble* any perceptible model in determinate ways. The single form of argument that rules out both ideas is one that employs this conception of theoretical representation.

Specific Difference and "Relative Ideas"

Having presented the case that it is Hume's understanding of theoretical representation that unites these two arguments, I will now turn to an alternative understanding of the conclusion of Hume's argument concerning external existence that also seeks to make use of the notion of theoretical representation, although in a way that is incompatible with the position that I have argued Hume occupies. As I understand that conclusion, it is quite strong: we can form *no idea* of anything other than perceptions. I have argued that Hume marshals his understanding of theoretical representation in support of this radical conclusion, and that that understanding requires a two-part interpretive thesis. Hume holds that one can use a perceptible model to represent a theoretical-explanatory entity if and only if one can specify the determinate ways in which the posited entity both differs from and resembles that perceptible model. This interpretive thesis can be objected to from at least two directions. One might object to that thesis on the grounds that Hume does not hold that we can represent theoretical entities at all. My hope is that the previous chapters have gone some way towards undermining objections from that direction (because Hume's explicit statements about the science of human nature indicate the opposite, and because his actual procedure in carrying out that science does the same). Still, one might also object to that thesis from the opposite direction: that Hume does allow for theoretical representation, but does not require that we use perceptible models in constructing these. That is, one might hold that Hume allows for the representation of objects that either differ from or resemble perceptions *in no way*. That thesis is at the core of the "New Hume" approach to the *Treatise*, and one of the textual pillars on which it rests its case is the paragraph that immediately follows Hume's announcement (and stylistic emphasis) of his conclusion that we cannot represent anything but our own perceptions.[4]

> 'tis impossible for us so much as to conceive or form an idea of any thing specifically different from ideas and impressions. Let us fix our attention out of ourselves as much as possible: Let us chace our imagination to the heavens, or to the utmost limits of the universe; we never really advance a step beyond ourselves, nor can conceive any kind of existence, but those perceptions, which have appear'd in that narrow

> compass. This is the universe of the imagination, nor have we any idea but what is there produc'd.
>
> The farthest we can go towards a conception of external objects, when suppos'd *specifically* different from our perceptions, is to form a relative idea of them, without pretending to comprehend the related objects.
>
> T 1.2.6.8–9; SBN 67–8

To form an idea of an object that is *specifically* different from our perceptions is impossible for the reason that we have already seen: doing so would require forming an idea of something that *in no way* resembles anything that we have observed. As I understand this passage, what Hume does consider for a moment is the possibility of forming "a relative idea of them," but he immediately makes clear here that "relative idea" is a misnomer. A relative idea is not really a kind of idea at all, because it does not even so much as "pretend to comprehend" its object. That is, a relative idea does not really represent any object at all. What does it do? Well, from what Hume says here, it seems that a relative idea does, somehow, at least gesture at "the related object." A relative idea is the idea of something being *related* to something else in some way, but without that something's being represented as having any specific intrinsic properties.[5] That is, a relative idea is a *via negativa*, and Hume's point here is that as far "external existence" is concerned, if one means by this something *specifically different* from perceptions, then since there is *ex hypothesi* no perceptible model to provide content to this idea, the most that one can do is conceive of it as *not* a perception, which amounts to *not conceiving of it at all.*

That is my take on this passage, but as I said, what I want to do in this section is to critically consider an alternative to that reading. Galen Strawson has pursued the thesis that Hume here articulates an important contrast between ideas with comprehensible descriptive content and "relative ideas," which do not have any such content but refer to a determinate object nonetheless. It will be worth quoting Strawson's explication at some length.

> This is the farthest we can go. But a relative idea is not no idea at all. True, we can't 'comprehend' external objects in any way on the terms of the theory of ideas. We can't in any way positively grasp or encompass (consider the etymology of 'comprehend') their real nature in thought. We can't form any positively descriptively contentful conception of their nature. We can't form any positively descriptively contentful conception of their nature of a sort that could count as a genuine comprehending or taking hold (or *Begriff*) of them in thought, because according to the theory of ideas
>
> (1) we can form a positively contentful conception of something only out of impressions-and-ideas material,

i.e. impressions material—for all ideas are in turn formed only by a kind of copying and combining of impressions material. But we've just supposed that

(2) external objects are entirely different in species or kind from impressions and ideas,

and Hume takes it that it follows from (2) that

(3) any impressions-and-ideas material we have is useless, in the attempt to form a positively descriptively contentful conception of the nature of external objects,

where these are supposed to be specifically different from perceptions. One might as well try to get a congenitally blind person to form a positively contentful conception of particular colours by giving them shapes to feel.

But although we can't form any idea of external objects that counts as positively contentful on the terms of the theory of ideas, we can still form a relative idea of such objects. It's a merely relative idea because we can't in any way conceive of or positively descriptively represent the nature of an external object (when it's supposed specifically different from perceptions). We can conceive it only indirectly. We may for example conceive it as something that stands in a certain relation (the relation of cause) to our perceptions

(cf. 84/1.3.5.2)[6]

The last line here is especially important because Strawson holds that it is this very notion of that-which-is-the-cause-of-my-impressions that he will use to argue that Hume is a realist about causal laws and the external world. His thought is that while we may not be able to *describe* the external objects that are the causes of our impressions, that phrase itself is enough to secure *reference* to such objects, which is all one needs to be a believer in them. Of course, this notion is also crucially important to the current study insofar as it is in direct contrast to the thesis that for Hume nothing can be represented that is not either directly observable or an analogical extension of some perceptible model. In fact, the situation is even more dire than that. If Strawson were right that Hume is both a causal realist *and a skeptic* (because while external objects can be referred to, they cannot be described), not only would it be the case that we *could* refer to undescribables, but it would also be the case that we could never *know* anything about such objects, and so both of the central tenets of the scientific realist interpretation that I have been defending would be false.

Fortunately, while the stakes are quite high here, I believe that a careful consideration of the case will show that both the exegetical and philosophical reasons for rejecting Strawson's interpretation are substantial.[7]

To begin, if the referent of a relative idea is fixed entirely independently of any description that we can give of that referent, it seems worth asking after what account of reference Strawson has in mind for doing this job. Farther along in his discussion of relative ideas, he offers a hint of what he is thinking.

> When Hume says that something X is unintelligible he means the same as Locke. [. . .] He doesn't mean that the term 'X' can't be supposed to refer to X, or that the notion of X is an incoherent one—any more than Locke supposed that the notion of the unknown real essence of gold was an entirely incoherent one, or that the word 'gold' could not 'carry reference' to the real essence of gold, despite the fact that he thought that this real essence might be forever undiscoverable by us.[8]

In a footnote to this passage Strawson cites Locke 1689–1700: 3.10.17–19, and Mackie "Locke's Anticipation of Kripke." He also makes reference to Hume's "Lockean" account of reference frequently throughout the rest of his study.[9] What Mackie finds in Locke is Kripke's notion of rigid designation, and since we are in search of a semantic theory that can provide the kind of reference that Strawson needs for Hume, it will be worth pausing to see if Kripke's can do the trick, especially since rigid designation itself was introduced specifically within the context of Kripke's articulation of just such a theory.

Kripke's theory of reference has two parts. The referent of a term is initially fixed through a baptismal event wherein a name is bestowed upon some object or natural kind. Subsequently, uses of that name refer to that object or natural kind in virtue of being linked to that baptismal event via some causal chain. So, for example, a child is conceived, the parents of the child pronounce that the child's name is "Violet Madeline", that pronouncement is the proximate cause of the doctor's uses of "Violet Madeline" and the distal cause of others' use of that same name. In this way the name "Violet Madeline" comes to refer to the person Violet Madeline. Alternatively, in the case of a natural kind of term, we encounter a certain liquid in our lakes and rivers and in a similar baptismal event dub this "water". This event is the proximate or distal cause of subsequent uses of "water" and in this way those uses come to refer to whatsoever it is that, in fact, fills our lakes and rivers (in our case, H_2O).

Now, one problem with this account is that Kripke says very little about how the baptismal event wherein the reference of a term is affixed to some object or natural kind is supposed to work.[10] One detail that Kripke gives, however, is that in general the semantic referent of a term is fixed by what he calls the speaker's reference, and that speaker's reference is fixed by the intention of a speaker to refer to a certain object or natural kind.

> So, we may tentatively define the speaker's referent of a designator to be that object which the speaker wishes to talk about, on a given occasion,

> and believes fulfills the conditions for being the semantic referent of the designator. He uses the designator with the intention of making an assertion about the object in question (which may not really be the semantic referent, if the speaker's belief that it fulfills the appropriate semantic conditions is in error).[11]

The problem that this move poses for *Kripke* is that it means that without saying more, his account of semantic reference presupposes an account of speaker's reference that remains utterly mysterious. That is, what Kripke attempts to offer is an explanation of how the reference of a name is fixed. After some machinations, that account boils down to an appeal to the intention of a speaker. If that appeal is to be *explanatory*, however, it must be the case that the reference of the intention of a speaker is better understood than the reference of a name, but there is little reason to think that it is. That is, Kripke's explanation of semantic reference in terms of speaker reference requires that we understand speaker reference, but since it is a *theory of reference* that Kripke is after, what he offers seems to make no progress at all. Of course, one is free to hold that speaker reference *is* better understood than semantic reference, although *Hume* rejects what is likely the most straightforward reason for holding that thesis, namely, that perceptions have a kind of primitive intentionality.[12]

The problem for *Strawson*, then, is that Kripke's semantic theory can provide no help in accounting for how a relative idea comes to refer to whatsoever-is-the-cause-of-my-impressions. The thought here was that just as for Kripke a name comes to refer to whatever object or natural kind to which it is first affixed, no matter what description we might give of this object or natural kind, so would relative ideas come to refer to whatever object or objects are the causes of impressions regardless of whether we can give any description of them at all. But if that move is going to work, then there must be some story of how such a referring term or idea comes to refer to its hidden object. For Kripke, a referring term gains its referent in virtue of the intention of a speaker to refer to some objects (and this mental state remains unexplained). But since what Strawson needs is an account of the reference of such mental states themselves, Kripke can be of no help to him.

How, then, is the reference of a "relative idea" fixed? Well, returning to Strawson's own picture what we find is actually more along the lines of the *descriptivist* account to which Kripke contrasts his own semantic theory. Rather than a reference-fixing event that determines the referent of a term independent of any description of it, what we find is a particular description being the *sole means* of referring to external objects. Namely,

> We may for example conceive it as something that stands in a certain relation (the relation of cause) to our perceptions

Our description of an external object as that which causes our perceptions, like the crude descriptivist's designation of Socrates as the teacher of Plato, picks out *whatsoever* happens to be that cause. This mode of designation proves to be a problem for the descriptivist because, as Kripke points out, we can be wrong about arbitrarily many things that we believe of the historical figure Socrates including that he was, in fact, the teacher of Plato, while our term "Socrates" would still refer to that historical figure. That won't be a problem for Strawson, though, because the description that fixes the reference of relative ideas is meant to be error-proof in just this way. Its reference *is* entirely fixed by that description, and unlike "Socrates", we cannot even in principle discover that we were wrong that the reference of the phrase "that which causes our impressions" did not cause our impressions.[13]

So, what we need to do now is look at that description more closely. What is described is that which stands in a certain relation to our perceptions. A natural place to look to understand it, then, is with Hume's account of relations.

> The word Relation is commonly used in two senses considerably different from each other. Either for that quality, by which two ideas are connected together in the imagination, and the one naturally introduces the other, after the manner above-explained; or for that particular circumstance, in which, even upon the arbitrary union of two ideas in the fancy, we may think proper to compare them.
>
> T 1.1.5.1; SBN 13

There are two senses of "relation": the common one and the philosophical one. In the common use "relation" refers to naturally occurring connections of ideas; in the philosophical one it refers to arbitrary relations of ideas in the fancy. Notice that in both cases, though, the *relata* of relations are ideas. Neither sense of "relation" allows for perceptions to be related to anything other than perceptions, and so the notion of "something standing in a certain relation to our perceptions" is a kind of category mistake. The *relata* of relations are ideas. So nothing that is not an idea can stand in a relation.

Of course, there is good reason to think that Hume has a bit more in mind here, and specifically, that his explication of relations in terms of ideas actually relies on an oblique reference to the *objects* of those ideas. In that case, what Hume writes here does not imply that relations can only hold between ideas, but rather that relations are what are represented by ideas when those ideas are themselves "connected together in the imagination," etc. So, for example, when one associates two ideas via a relation of spatial contiguity (i.e., one forms a mental image by placing two ideas next to each other), one thereby pictures the spatial relation of the objects of those ideas. That is a more plausible take on the nature of relations, and one that makes good sense of the fact that we have seen Hume repeatedly write as if relations do hold among not only the ideas that are the vehicles of our representing, but also among those objects thereby pictured.

Notice, though, that this latter reading of Hume's account of relations gives no succor to Strawson's attempt to cast relative ideas as relating our perceptions to that which is their undescribable cause. Even understanding Hume as holding that a relation is that which is represented by ideas, a relation can still only hold between items that are themselves the object of ideas *with descriptive content*. That is, if a relation is that which is represented by some complex idea, relations can only exist between items that are themselves the objects of some ideas. Since, *ex hypothesi*, the object of a relative idea is not represented by any idea, it is not so much as a candidate *relatum* of any represented relation.

The details of Hume's two definitions of "cause"—the specific relation to which Strawson appeals—supports this understanding.

> We may define a cause to be "An object precedent and contiguous to another, and where all the objects resembling the former are plac'd in like relations of precedency and contiguity to those objects, that resemble the latter." If this definition be esteem'd defective, because drawn from objects foreign to the cause, we may substitute this other definition in its place, *viz.* "A cause is an object precedent and contiguous to another, and so united with it, that the idea of the one determines the mind to form the idea of the other, and the impression of the one to form a more lively idea of the other."
>
> T 1.3.14.31; SBN 170

As Garrett argues, just as we can understand Hume's explication of relations in general as appealing to either or both of the features of the ideas that represent relations or the relations represented by those ideas, each of these definitions can be understood as a way of indicating the members of the revival set of the general term "cause".[14] The first specifies the set by appealing to the common features of the objects represented by its member ideas, whereas the second specifies it by appealing to the features common to the ideas themselves. In either case, though, the causal relation is one that holds between ideas or the objects of ideas *with descriptive content*. To claim of something that it is the cause of something else, it must be at least possible for the former to be a member of the revival set of "causation", and that is, *ex hypothesi*, not possible for the object of Strawson's "relative ideas".

Another passage from 1.1.5 is also illustrative here. Since a "relative idea" is supposed to be the most that we can form once we have supposed that external objects are specifically *different* from perceptions, consider Hume's account of difference.

> It might be naturally expected that I shou'd join *difference* to the other relations. But that I consider rather as a negation of relation, than as any thing

> real or positive. Difference is of two kinds as oppos'd either to identity or resemblance. The first is called difference of *number*; the other of *kind*.
>
> T 1.1.5.10; SBN 15

Difference is not a relation, but the negation of a relation, either the negation of the relation of identity or of resemblance. So, in saying of two perceptions (or objects) that they are different, we might mean either that those two perceptions are not numerically identical to each other, or that they do not resemble each other. In both cases, though, it must be possible to represent the relation at hand in order to then deny that such a relation holds. That is, Hume's point here is that there is no relation that is simply the relation of two items' being different *simpliciter*. To represent two items as being different, one must first represent a way in which they might have been the same—numerical identity or qualitative resemblance—and then negate this representation, so to speak.

That way of understanding difference, however, rules out precisely Strawson's understanding of a relative idea as referring to something specifically different from perceptions because that understanding casts relative ideas as representing a relation of difference *simpliciter*. A relative idea as Strawson understands it, would not be the negation of a representation of a specific relation between one perception and another, but rather would be the representation of that which is not a perception at all. Unlike the cases that Hume describes as cases of difference, there is no determinate contrast on offer in Strawson's case. While we can contrast subsets of perceptions with one another—e.g., ideas resemble each other in determinate ways, whereas impressions do not resemble ideas in those ways—since we cannot represent the necessary contrast class to "perceptions", we cannot represent perceptions as related to anything other than perceptions, and so also cannot negate this representation, i.e., represent anything as different from a perception. Again, relations can only hold between perceptions or the representational objects of perceptions, so only perceptions or the representational objects of perceptions can differ, i.e., be the *relata* of a negated relation.

Furthermore, even if we could form an idea the entire content of which was an object that is not identical to any perception and that resembled perception in no way, notice that even in that case, the relation that would constitute that relative idea would not be "any thing real or positive." A relative idea neither comprehends its object nor even so much as represents its object as standing in any real relation to anything. To have a relative idea is not to have any idea at all. What then, is our idea of an external object, if not the idea of something specifically different from perceptions? The relative idea passage continues:

> Generally speaking we do not suppose them [external existences] specifically different; but only attribute to them different relations, connexions, and durations. But of this more fully hereafter.
>
> T 1.2.6.9; SBN 68

Here the idiom of relation recurs, but now not with respect to the limits of the notion of specific difference, but instead in contrast to that notion as the key to forming a notion of external existence as something that is *not* specifically different from perceptions. The final sentence, of course, is a reference forward to 1.4.2, where Hume will detail precisely how we form a notion of what he calls body after all, namely by changing the relations, connections, and durations that structure our complex ideas, i.e., by using our own perceptions as a perceptible model for that notion, and then specifying the determinate ways in which the theoretical posit—body—differs from and resembles that model, i.e., by relating that posit to that model via "relations" other than merely difference alone.

Before moving on to our consideration of 1.4.2, though, it will be worth addressing one final piece of textual evidence from Strawson. Farther along in his discussion of specific difference, Strawson presents what appears to be a smoking-gun set of passages.

> Whatever difference we may suppose betwixt . . . [a perception and an external object taken to be specifically different from a perception], 'tis still incomprehensible to us; and we are oblig'd either to conceive an external object merely as a relation without a relative, or to make it the very same [as] a perception or impression.
>
> (241/1.4.5.19)

Five lines later:

> we may suppose, but never can conceive a specific difference betwixt an object and an impression.[15]

Presented this way, it certainly does look like Hume delineates two ways to understand "external existence" qua specifically different from a perception. On the one hand, we can do as I have been suggesting and "make it the very same with a perception or impression." On the other, we can do what Strawson suggests and, "conceive an external object merely as a relation without a relative." Hume appears to then confirm the latter as a genuine possibility in the second passage in claiming that we can "suppose [. . .] a specific difference betwixt an object and an impression". Strawson is right that presented this way, this looks to be as bad as it can get for his opponents. Of course, *Hume* does not present these passages in this way. Hume prefaces the first passage by prompting his reader to,

> remember, that as every idea is deriv'd from a preceding perception, 'tis impossible our idea of a perception, and that of an object or external existence can ever represent what are specifically different from each other.
>
> T 1.4.5.19; SBN 241

That is about as explicit an endorsement of the claim that we cannot represent anything specifically different from our perceptions as one could hope for. Period. What then of Strawson's quotation from five lines down? Well, Strawson truncates that sentence—the period that ends his quote is not Hume's—and omits entirely the argument of which it is a part. Here is the full text.

> I say then, that since we may suppose, but never can conceive a specific difference betwixt an object and impression; any conclusion we form concerning the connexion and repugnance of impressions, will not be known certainly to be applicable to objects; but that on the other hand, whatever conclusions of this kind we form concerning objects, will most certainly be applicable to impressions. The reason is not difficult. As an object is suppos'd to be different from an impression, we cannot be sure, that the circumstance, upon which we found our reasoning, is common to both, supposing we form the reasoning upon the impression. 'Tis still possible, that the object may differ from it in that particular. But when we first form our reasoning concerning the object, 'tis beyond doubt, that the same reasoning must extend to the impression: And that because *the quality of the object, upon which the argument is founded, must at least be conceiv'd by the mind; and cou'd not be conceiv'd, unless it were common to an impression*; since we have no idea but what is deriv'd from that origin.
>
> T 1.4.5.20; SBN 241–2, emphasis mine

As the italicized portion of this passage shows, the text that Strawson cites, far from indicating that Hume thought that we can form a relative idea of an object, is in fact part of an argument that explicitly denies the possibility of any such thing. The reason that we can infer the qualities of an impression from the qualities of an object is precisely because *the only way* to form an idea of an object is if the qualities of the object are "common to an impression." That is, the only way to form an idea of an object is by taking that object to share the qualities of some impression, i.e., to employ a perceptible model in representing it, i.e., we cannot represent anything specifically different from a perception!

One final note before leaving Strawson.[16] It is part of Strawson's interpretation that while Hume does not find the idea of external existence intelligible, he nonetheless believes in such an existence.

> In fact Hume not only accepts that it is intelligible to suppose that there may exist things which are (E-)unintelligible to us. He also accepts that there actually are such things—that there is such a thing as the 'ultimate . . . perfectly inexplicable . . . cause' of our perceptions (84/1.3.5.2); that there are (utterly 'incomprehensible' and 'inconceivable') 'powers and forces, by which the [course of nature] is governed'
>
> (54/5.21)[17]

According to Strawson, Hume accepts that there exist such things that are unintelligible to us, and also *believes* in such things. Of course, Hume has an account of belief.

> 'Tis also evident, that *the idea of existence is nothing different from the idea of any object*, and that when after the simple conception of any thing we wou'd conceive it as existent, we in reality make no addition to or alteration on our first idea. Thus when we affirm, that God is existent, we simply form the idea of such a being, as he is represented to us; nor is the existence, which we attribute to him, conceiv'd by a particular idea, which we join to the idea of his other qualities, and can again separate and distinguish from them. But I go farther; and not content with asserting, *that the conception of the existence of any object is no addition to the simple conception of it*, I likewise maintain, that the belief of the existence joins no new ideas to those, which compose the idea of the object. When I think of God, when I think of him as existent, and when I believe him to be existent, my idea of him neither encreases nor diminishes.
>
> T 1.3.7.2; SBN 94, emphasis added

> An opinion, therefore, or belief may be most accurately defin'd, A lively idea related to or associated with a present impression.
>
> T 1.3.7.5; SBN 96

It is clear in both of these passages that belief in the existence of an object requires an *intelligible* idea of that thing, not just a relative one. In the first passage we see that, "the conception of the existence of any object is no addition to the simple conception of it." That claim implies that to believe in the existence of any object, we must first have a *conception* of it. Conception, of course, is what Strawson opposes to "supposition" in articulating the difference between intelligible ideas and merely relative ones. So, if it is conception, rather than merely supposition, that is required for belief, it is unclear in what a belief in an existence that is specifically different from our perceptions could consist.[18]

Strawson might be tempted to here rejoin that his notion of supposition is not opposed to conception *per se*, but only to *direct* conception, and that supposition is itself a kind of *indirect* conception. The second passage above, the definition of belief itself, rules this out, though. An indirect conception is one in which the idea at hand has whatever purely referential (i.e., non-descriptive) content that it does not by being related to any impression, but only in virtue of standing in some relation to an unknown and unintelligible object. What the definition of belief states, though, is that a belief is a lively idea *related to or associated with a present impression*. In the case of a relative idea, though, there is *ex hypothesi* no present impression that can serve this function. It is in virtue of an idea's resembling and being caused by some impression that the liveliness of the impression is carried over to the idea, making the latter into a belief. But, relative ideas do not resemble and are not caused by any impression, and so this transfer of liveliness cannot occur, and therefore no belief in the object of a relative idea is possible.

Finally, notice that Hume bundles all of the points that we have made here—concerning relations, causation, and belief into a single passage that appears to explicitly rule out precisely the kind of belief-in-something-we-know-not-what that Strawson advocates.

> *I shall* add as a fourth corrollary, that we can never have reason to believe that any object exists, of which we cannot form an idea. For as all our reasonings concerning existence are deriv'd from causation, and as all our reasonings concerning causation are deriv'd from the experienc'd conjunction of objects, not from any reasoning or reflection, the same experience must give us a notion of these objects, and must remove all mystery from our conclusions. This is so evident, that 'twou'd scarce have merited our attention, were it not to obviate certain objections of this kind, which might arise against the following reasonings concerning *matter* and *substance*. I need not observe, that a full knowledge of the object is not requisite, but only of those qualities of it, which we believe to exist.
>
> T 1.3.14.36; SBN 171

When it comes to representing and believing in matter or substance, there can be no reason to believe in that of which we cannot form an idea because all of our reasoning about the existence of substance will proceed via reasoning about causation—as we discussed in previous chapters, the idea of substance is the idea of certain properties united by a single causal explanation—and reasoning about causation can only proceed where experience has afforded the resources for representing those objects that are thereby related. Also worth noting is the final sentence of this passage, where Hume himself notes that in representing and endorsing the existence of a theoretical entity, one need not represent *every* quality of such a posit, but rather only those that are specifically required to do the explanatory work at hand. That is, so long as a theoretical posit resembles some perceptible model in certain determinate ways, it can be left open what further features that which is posited might have. So, while we cannot, as Strawson contends, represent that which causes our perceptions as a mere *via negativa*, causal explanations are also not limited merely to the contents of direct experience. Once again, we find that Hume occupies a middle ground between the rationalist mechanical philosophers that preceded him and the inductivist anti-realist that he has been taken to be.

In sum, Strawson's interpretation of relative ideas fails on a number of fronts. Hume cannot piggyback on a Lockean/Kripkean theory of reference because such a theory itself presupposes a theory of reference for mental items that Strawson's Hume cannot provide. Hume's account of relations likewise prohibits him from using a non-Kripkean descriptivist theory of reference to account for the content of relative ideas because the description, "that-which-causes-our-perceptions," does not describe any possible relation. Furthermore, the textual evidence that Strawson provides for his reading, when examined in its proper context, actually speaks to precisely the opposite thesis: that relative ideas are no ideas at all. Other textual evidence points to the same conclusion. Therefore, I conclude that Hume's thesis is that we can form no idea of any kind of anything that is specifically different from our perceptions, i.e., that we can form no idea of anything that does not resemble our perceptions in some way, i.e. that all representation requires either direct observation or the deployment of a perceptible model.[19]

Before moving on, though, it is important to notice that while I have been resisting Strawson's claim that Hume believes in the existence of something specifically different from perceptions, *this* claim of Strawson's is actually only an instantiation of a more general claim of his with which I am in agreement. That more general claim is that,

> Hume never really questions the idea that there is something like causal power in reality, i.e. *something about reality in virtue of which reality is regular in the way it is.*[20]

This notion, which is the same as the demand for an *explanation* of the empirically discovered universal generalizations and regularities that govern experience—lies at the very core of my interpretation of Hume's pursuit of the science of human nature. Where my interpretation differs from Strawson's is that Strawson holds that these powers could only be the powers of objects that are specifically different from perceptions and therefore that, while we can refer to them using a "relative idea," they are entirely unknowable to us.[21] By contrast, I hold that what explains the regularities of perception, where such explanation is possible, is the existence of a particular kind of substance, our understanding of which is modeled on perceptions themselves (and so such substances cannot be specifically different from those perceptions). It is important to note here from whence this difference arises. Strawson understands Hume's commitment to there being something that explains the regularities discovered in experience as also committing him to the existence of something specifically different from perceptions. I accept the commitment to explanation, but not the commitment to external existence. Hume pursues the science of human nature exactly as far as the principles of that science allow:

> Let us fix our attention out of ourselves as much as possible: Let us chace our imagination to the heavens, or to the utmost limits of the

> universe; we never really advance a step beyond ourselves, nor can conceive any kind of existence, but those perceptions, which have appear'd in that narrow compass. This is the universe of the imagination, nor have we any idea but what is there produc'd.
>
> T 1.2.6.8; SBN 67–8

Because any theoretical-explanatory posit must be modeled on perceptions themselves, whatsoever we posit as an explanation of our experience, will always likewise be or be relevantly similar to a perception. Thus, it is theoretical-explanatory science itself that limits the scope of the representable, understandable, and knowable universe to the mind itself.

The Explanations of the Vulgar and of the False Philosophy

Having now completed our investigation into the conclusion from 1.2.6 that frames Hume's discussion of the idea of body in 1.4.2–5—the thesis that we cannot form an idea of anything specifically different from our perceptions—what I want to do in this section is walk through Hume's presentation of the failures of the vulgar and the false philosophers to explain the apparent constancy and coherence of our perceptions in order to note the role that his own brand of scientific realism plays in his understanding of them.

Hume begins 1.4.2 with an investigation into the source of our idea of body, and quickly discovers that this idea is not a product of either the senses or reason, but rather of the imagination. (This discovery will prove significant farther along insofar as, as we have previously noted, the products of the imagination have substantially less claim on truth than do those of the senses and reason.) That discovery prompts Hume to investigate the cause of the imagination's producing this idea, and he almost immediately reports that,

> After a little examination, we shall find, that all those objects, to which we attribute a continu'd existence, have a peculiar *constancy*, which distinguishes them from the impressions, whose existence depends upon our perception. These mountains, and houses, and trees, which like at present under my eye, have always appear'd to me in the same order; and when I lost sight of them by shutting my eyes or turning my head, I soon after find them return upon me without the least alteration.
>
> T 1.4.2.18; SBN 194

Part of what prompts the imagination to formulate the idea of body is certain regularities of experience: that I have perceptions of these mountains, and houses, and trees all in the same order, over and over again through a variety of other circumstances calls out for an explanation, and it is the imagination that answers this call. Hume makes it explicit that the idea of body is an attempt at this form of explanation—explaining the fact that the particulars conform, to the extent that they do, to certain empirical generalities via an appeal to their nature—in the following paragraph where he notices that it cannot be this constancy alone that is operative here.

> This constancy is not so perfect as not to admit of very considerable exceptions. Bodies often change their position and qualities, and after a little absence or interruption may become hardly knowable. But here 'tis observable that even in these changes they preserve a *coherence*, and have a regular dependence on each other; which is the foundation of a kind of reasoning from causation, and produces the opinion of their continu'd existence.
>
> T 1.4.2.19; SBN 195

While the constancy of our perceptions admits of exceptions, much of this is mitigated by a kind of meta-regularity. Hume's example is that while his fire is not in the same state that was when he earlier left his room, he has observed a corresponding change in the states of his fire repeatedly in the past. That is, while not everything in perception remains constant, the changes that perceptions undergo are themselves discoverable to be quite regular. As I have argued in the past three chapters, it is exactly such regularities that call for an explanation by reason, and in the final clause of this passage we see that Hume cites just such a demand as what first prompts the formation of the idea of body. The coherence of our perceptions, "is the foundation of a kind of reasoning from causation, and produces the opinion of their continu'd existence." Reasoning from causation is exactly what the understanding of Hume's science of human nature that I have been defending predicts: a regularity is discovered, and reason combines with sensation to form a representation of the substance (a general term with a causal uniting principle as its foundation) that explains these regularities.

Notice, though, that Hume hedges here. He reports not that coherence prompts a bit of reasoning from causation, but rather that it prompts *a kind* of reasoning. Two paragraphs later, he confirms his hesitation.

> But tho' this conclusion from the coherence of appearances may seem to be of the same nature with our reasonings concerning causes and effects; as being deriv'd from custom, and regulated by past experience; we shall find upon examination, that they are at the bottom considerably different from each other, and that this inference arises from the understanding, and from custom in an indirect and oblique manner.
>
> T 1.4.2.21; SBN 197

What appears to be an inference to the best explanation (reasoning about causation) in the case of the coherence of our perceptions is not so because while our reasoning about causation is "deriv'd from custom and regulated by past experience," the formulation of the idea of body is not. To understand this claim, we need to know what it means to be "deriv'd from custom and regulated by past experience." If one understands Hume as aiming at merely discovering the empirical regularities that govern the mind, this phrase suggests that whatever new ideas we can legitimately produce must be "more of the same," i.e., must not do more than repeat and rearrange what has already been experienced. On that reading, the difference between causal reasoning and the idea of body is that the latter attempts to represent something novel, whereas the former merely notices connections between what has already been experienced.

I have already rejected this impoverished understanding of Hume in favor of one that makes theoretical representations with novel content possible via analogical extensions of some perceptible model. Accordingly, as I understand the demand that reasoning about causation be "deriv'd from custom and regulated by past experience," it is precisely the condition that all theoretical representation contains a mixture of reason and impressions, and that experience be used both to discover the regularities that demand explanations of this kind, and that any such theoretical-explanatory posits depend for their legitimacy on their continued coherence with experience. The problem with the idea of body is that it is not a mixture of *reason* and impressions at all.[22]

> But as all reasoning concerning matters of fact arises only from custom, and custom can only be the effect of repeated perceptions, the extending of custom and reasoning beyond the perceptions can never be the direct and natural effect of the constant repetition and connexion, but must arise from the co-operation of some other principles.
>
> T 1.4.2.21; SBN 198

Merely being the product of the imagination instead of reason does not by itself undermine the legitimacy of the idea of body, but additionally requires an argument that the imagination is an unreliable source of explanation, and therefore truth. This is precisely what Hume offers in the next paragraph.

> I have already observ'd, in examining the foundation of mathematics, that the imagination, when set into any train of thinking, is apt to continue, even when its object fails it, and like a galley put in motion by the oars, carries on its course without any new impulse. [. . .] Objects have a certain coherence even as they appear to our senses; but this coherence is much greater and more uniform, if we suppose the objects to have a

> continu'd existence; and as the mind is once in the train of observing an uniformity among objects, it naturally continues, till it renders the uniformity as compleat as possible.
>
> T 1.4.2.22; SBN 198

The problem with the imagination is not that it posits novel content *per se*, but rather that it does so on illicit grounds. Specifically, the imagination is highly susceptible to *confirmation bias*. Once the imagination comes to believe that a certain regularity exists, it "renders the uniformity as compleat as possible," by paying more attention to instances that confirm the existence of that regularity, and by ignoring disconfirming evidence. As Hume later notes,

> I cannot conceive how such trivial qualities of the fancy, conducted by false suppositions, can ever lead to any solid and rational system.
>
> T 1.4.2.56; SBN 217

It is not because the imagination employs theoretical-explanatory posits that its formulation of the idea of body is illegitimate, but rather because in making that posit it is, "conducted by *false suppositions*." That Hume finds the imagination to be unreliable because it is susceptible to bias and false belief indicates that a faculty that was not susceptible in these ways could, "lead to any solid and rational system," i.e., legitimately and successfully provide theoretical explanations that moved beyond experience. Of course, Hume takes reason to be just such a faculty.

> Now upon that supposition, 'tis a false opinion that any of our objects, or perceptions, are identically the same after an interruption; and consequently the opinion of their identity can never arise from reason, but must arise from the imagination.
>
> T 1.4.2.43; SBN 209

Notice how Hume's argument in this brief passage proceeds. Because it is false that any of our perceptions are identically the same after an interruption, *consequently the opinion of their identity can never arise from reason*. A false opinion can never arise from reason. Reason is a faculty that guarantees the truth of its conclusions. That sentiment has to be an exaggeration, given that Hume is explicit that reasoning concerning causation is fallible, but what it nonetheless captures is the contrast that Hume means to draw between reason and the imagination. The former is in general extremely reliable, whereas the latter is not. So, again, the problem with the formulation of the idea of body is not that theoretical-explanatory posits are illegitimate *per se*, but rather that this particular theoretical-explanatory posit is the product of an unreliable faculty.[23]

Of course, even unreliable faculties can occasionally produce true ideas. Thus, Hume rightly proposes that,

> 'Twill first be proper to observe a few of those experiments, which convince us, that our perceptions are not possest of any independent existence.
>
> T 1.4.2.45; SBN 210

The hypothesis of the vulgar, that our perceptions exist independently of our perceiving them is not only the product of an unreliable faculty, but can also be demonstrated to contradict the empirical evidence.

> When we press one eye with a finger, we immediately perceive all the objects to become double, and one half of them to be remov'd from the common and natural position.
>
> T 1.4.2.45; SBN 210–11

> This opinion is confirm'd by the seeming increase and diminution of objects, according to their distance;[24] by the apparent alterations in their figure; by the changes in their colour and other qualities from our sickness and distempers; and by an infinite number of other experiments of the same kind; from all of which we learn, that our sensible perceptions are not possest of any distinct or independent existence.
>
> T 1.4.2.45; SBN 211

Again, the reason to reject the idea of body is not that it is a theoretical-explanatory posit, but rather that it is a theoretical-explanatory posit that (a) is the product of an unreliable faculty, and (b) that it is demonstrably false! If theoretical-explanatory posits were *per se* illegitimate, then there would be no need to demonstrate their incompatibility with experience. One could rule out their use on purely philosophical grounds. That, however, is not what we find Hume doing.

It is also worth noting here that the vulgar's attempt to use a theoretical posit to explain the constancy and coherence of our perceptions, for all its faults, does accord with the understanding of Hume's conception of theoretical explanation.[25] That is, one thing that the vulgar get right is using observed perceptions as a model for explaining the regularities that are discovered in experience, and accompanying this model with a commentary specifying more or less determinate ways that the posited substance resembles and differs from it. (The vulgar explain the observed, and unobserved(!), regularities of experience by taking these to *really* consist of encounters with a kind of perception, modeled on observed perceptions, that continue to exist when we do not perceive them and that exist distinctly from our perception of them.)[26] The imagination exaggerates the observed regularities, and produces a hypothesis that directly contradicts empirical evidence, and so its *results* cannot be accepted, but at least this much of its process is in line with good scientific practice. As we will see in a moment, not even this much can be said of the explanations of the false philosophers, and so it is that Hume treats them with significantly more disdain.

Before moving on from the vulgar, though, I want to pause to address a few possible objections. My claim has been that Hume rejects the explanations of the vulgar in part because he takes their *explicanda* to be illusory. The constancy and coherence of our perceptions are greatly exaggerated by the illicit activity of the imagination. One line of objection to this thesis would be to resist the claim that Hume treats the purported constancy and coherence of our perceptions similarly, saving one at the expense of the other. Furthermore, this line can be pursued along both philosophical and exegetical dimensions.[27] Exegetically, it might appear that Hume distinguishes between the effects of the supposed constancy and coherence of our perceptions, and emphasizes that the latter plays a smaller role in the vulgar's conception of body than does the former.

> But whatever force we may ascribe to this principle, I am afraid 'tis too weak to support alone so vast an edifice, as is that of the continu'd existence of all external bodies; and that we must join the *constancy* of their appearance to the *coherence*, in order to give a satisfactory account of that opinion.
>
> T 1.4.2.23; SBN 198–9

Philosophically, one might well wonder whether constancy is even the sort of thing that can come in degrees: for our perceptions to be constant is just for the perception that we have immediately before an interruption to exactly resemble the one that we have immediately following that interruption. Such are the circumstances in which many of our perceptions appear, as in the examples that Hume cites.

> Those mountains, and houses, and trees, which lie at present under my eye, have always appear'd to me in the same order; and when I lose sight of them by shutting my eyes or turning my head, I soon after find them return upon me without the least alteration. My bed and table, my books and papers, present themselves in the same uniform manner, and change not upon account of any interruption in my seeing or perceiving them.
>
> T 1.4.2.18; SBN 194–5

These are clearly examples of the constancy of our perceptions, and equally clear is that they do not admit of degree. Nonetheless, I maintain that Hume takes the constancy of our perceptions to be a merely apparent regularity. By way of illustration, consider again Hume's introduction of the exact resemblance thesis from 1.1.

> The first circumstance, that strikes my eye, is the great resemblance betwixt our impressions and ideas in every other particular, except their degree of force and vivacity. The one seem to be in a manner the reflexion of the other; so that all the perceptions of the mind are double, and appear both as impressions and ideas. When I shut my eyes and think of my chamber, the ideas I form are exact representations of the impressions I felt; nor is there any circumstance of the one, which is not to be found in the other. In running over my other perceptions, I find still the same resemblance and representation. Ideas and impressions appear always to correspond to each other.
>
> T 1.1.1.3; SBN 2–3

Here Hume presents what appears to be a genuinely *universal* regularity: "ideas and impressions appear *always* to correspond to each other." As is familiar, though, Hume immediately walks the strength of this thesis back.

> Upon a more accurate survey I find I have been carried away too far by the first appearance [. . .] I observe, that many of our complex ideas never had impressions, that corresponded to them, and that many of our complex impressions never are exactly copied in ideas. I can imagine to myself such a city as the *New Jerusalem*, whose pavement is gold and walls are rubies, tho' I never saw any such. I have seen *Paris*; but shall I affirm I can form such an idea of that city, as will perfectly represent all its streets and houses in their real and just proportions?
>
> T 1.1.1.4; SBN 3

Hume presents his findings here dialectically. He feigns observing an exact resemblance between all of his impressions and ideas to motivate the notion that there is a strong connection between experience and thought, but then corrects this naïve conception of that connection by revealing that the regularity discovered is merely apparent, and that understanding the underling regularity requires further investigation and nuance.

A very similar dialectic is in play in Hume's presentation of the constancy of our perceptions. He first cites examples of what appears to be a universal regularity of experience—mountains, houses, trees, his bed and desk, etc.—and then immediately notes that the apparent regularity at hand is not as wide in its scope as it first appears.

> This constancy, however, is not so perfect as not to admit of very considerable exceptions. Bodies often change their position and qualities, and after a little absence or interruption may become hardly knowable.
>
> T 1.4.2.19; SBN 195

While Hume does not deny that there are instances where a perception appearing before an interruption exactly resembles one appearing immediately after that interruption, he does notice that the occurrence of such pairs is not as frequent as it might at first appear. That supposition admits, "of *very considerable* exceptions," and, "bodies *often* change their position," and qualities," and "may become *hardly knowable*." So, while a particular instance of a pair of perceptions being constant does not admit of degree, the frequency with which we encounter such pairs does, and far from amounting to a universal regularity that demands explanation, it is often the case that perceptions are not at all constant.

What then of the weight that Hume rests on constancy over coherence? Well, notice that in the quotation above from 1.4.2.23 Hume does not actually claim that constancy is more important than coherence. What he claims is that coherence alone cannot support the edifice of the hypothesis that bodies have a continued existence. Neither, however, can constancy alone: "we must *join* the constancy of their appearance to the coherence, in order to give a satisfactory account." As Hume then goes on to give that account he does appear to focus on constancy alone as the source for the vulgar's confused notion of the continued existence of our perceptions, but this really just a shorthand for constancy *and coherence*.[28] As we have seen, Hume holds that the (relatively infrequent) constancy of our perceptions would not alone be enough to give rise to the idea of continued existence, and that this must be supplemented and supported by the supposed meta-regularity of their coherence. It is the paragraph immediately preceding his presentation of his account of the idea of continued existence where he states that constancy and coherence must be joined to form an adequate of that idea. Thus, I conclude that the textual and philosophical evidence points to a Hume's taking both the supposed constancy and coherence of our perceptions to be greatly exaggerated, and therefore neither to be phenomena that genuinely demand explanation.

To summarize: when Hume describes the process whereby the vulgar come by the idea of body, he does so in a way that portrays them as pursuing a methodology, which while ultimately illegitimate, is close to what my proposed understanding of the science of human nature proscribes. The vulgar discover what they take to be a universal regularity among our perceptions: their constancy and coherence. To explain this regularity, they posit an additional kind of perception that is not dependent and fleeting, but rather distinct and continued. They represent this posit by taking their own perceptions as a model, and specifying the ways that the posited entity both resembles and differs from that model. Finally, they endorse the reality of this posit, they take it is representing the substance of the world, by adopting a set of general terms—they add "object" to their vocabulary—that represents this posit as different in kind from previously observed perceptions. As described, there is nothing wrong with this procedure in the slightest, and performed correctly, should yield an explanatory and true representation of the world. Unfortunately for the vulgar, the observed regularity at which their explanation is aimed is not a genuine one, but rather the result of the exaggeration of the unreliable imagination, and their hypothesis stands refuted by "an infinite number of experiments." Again, though, it is not in entertaining a theoretical-explanatory posit *per se* that the vulgar go wrong, but in misconstruing the empirical evidence for that hypothesis.

That said, the proponents of what Hume calls the false philosophy do not fare nearly so well. Here is Hume's initial introduction of the threefold distinction between the systems of the vulgar, the false philosophy, and the true (the science of human nature, which it is worth nothing, receives credit here for achieving the highest degree of knowledge and *reason*).

> In considering this subject we may observe a gradation of three opinions, that rise above each other, according as the persons, who form them, acquire new degrees of reason and knowledge. These opinions are that of the vulgar, that of a false philosophy, and that of the true.
>
> T 1.4.3.9; SBN 222

Thus far in the current chapter, we have been focused on the failure of the vulgar to properly describe and explain the apparent constancy and coherence of our perceptions. As it turns out, Hume takes there to be a second way that that system fails as well. Because the explanatory scope of the vulgar is narrowly focused on just the apparent constancy and coherence of our perceptions, the representation that they use to explain it does not contain the resources for also accounting for its own powers. That is, the explanation of the vulgar, in addition to being false, is also inadequate to the broader explanatory demands that human nature itself requires.

> The persons, who entertain this opinion concerning the identity of our resembling perceptions, are in in general all the unthinking and unphilosophical part of mankind, (that is, all of us, at one time or other) and consequently such as suppose their perceptions to be their only objects, and never think of a double existence internal and external, representing and represented.
>
> T 1.4.2.36; SBN 204

Drawing the distinction between a represent*ing* and that which is represent*ed* appears to allow the false philosophy to correct the mistakes that it discovers in the language of the vulgar by recasting the confounded perceptions represented by the vulgar as an amalgam of represent*ed* material substances and represent*ing* perceptions.

Notice the form that this purported explanation takes. Having observed certain empirical regularities in the behavior of their perceptions, the vulgar seek to explain these via forming a certain picture of the mind (as consisting of perceptions that continue to exist even when they are not perceived). That picture proves untenable, and so a potential successor picture is proposed: the false philosophy. That successor purports to explain both the original phenomena—the apparent constancy and coherence of our perceptions—as well as the apparent success and failure of the predecessor scheme. As the false philosopher sees it, the system of the vulgar worked to the extent that it did because there is *something* that exists even when we do not perceive it, but it ultimately fails because it confounds this represented something (material substance) with that which represents it (perceptions). Both explanations are ontologically committing: they both rely on explaining the relevant empirical generalities as resulting from the nature, essence, and powers of certain posited entities. The vulgar take our perceptions to exist independently of being perceived, and the false philosophy posits a material substance that does so.

So, Hume appears to cast both the language of the vulgar and that of the false philosophy as attempts to picture the essence, nature, and powers of the mind. We have seen why the vulgar fail to achieve this end, and now we must turn to the failure of that false philosophy, which will provide further confirmation that Hume subscribes to the kind of scientific realism that we have been attributing to him. Specifically, the false philosophy suffers from four insuperable deficiencies:

(1) it is not the product of *reason*,
(2) it seeks to explain what is in fact an *ultimate principle* of the human mind,
(3) its theoretical posit has no *perceptible model* and therefore relies on an abuse of language, and
(4) its apparent explanatory power is in fact entirely *parasitic* on that of the system of the vulgar.

We can begin with the first two of these failures, which Hume understands as intimately related. To do that, we begin where the false philosophy does: with the failure of the system of the vulgar.

> 'Tis natural for men, in their common and careless way of thinking, to imagine they perceive a connexion betwixt such objects as they have constantly found united together; and because custom has render'd it difficult to separate the ideas, they are apt to fancy such a separation to be in itself impossible and absurd. But philosophers, who abstract from the effects of custom, and compare the ideas of objects, immediately perceive the falshood of these vulgar sentiments, and discover that there is no known connexion among objects. Every different object appears to them entirely distinct and separate; and they perceive, that 'tis not from a view of the nature and qualities of objects we infer one from another, but only when in several instances we observe them to have been constantly conjoin'd. But these philosophers, *instead of*

> *drawing a just inference from this observation, and concluding*, that we have no idea of power or agency, separate from the mind, and belonging to causes; I say, instead of drawing this conclusion, they frequently search for the qualities, in which this agency consists, and *are displeas'd with every system, which their reason suggests to them, in order to explain it.*
>
> T 1.4.3.9; SBN 223, emphasis mine

Notice that the most straightforward conclusion to draw from the failure of the system of the vulgar would be that the apparent constancy and coherence of our perceptions is *explanatorily and causally basic*: "we have no idea of power or agency, separate from the mind, and belonging to causes". The proponents of the false philosophy violate what we saw earlier is one of Hume's most important norms regarding scientific explanation: that of restraining the intemperate desire to seek explanations where none are possible.

Next, notice that in the italicized portions of the passage Hume makes clear that the false philosophers are lead to this intemperance in part by neglecting the deliverances not just of the senses, but of *reason*, and more specifically by failing to draw the "inference from this observation." As we noted earlier, an idea is the product of reason just in case it is the result of the application of our inferential faculty where (as the numerous examples that we considered showed) that faculty must be taken to include inference to the best explanation. Hume's objection to the false philosophers here is that reason delivers such an idea here—"every system, which their reason suggests to them, in order to explain it"—but that the false philosophy ignores this conclusion and pursues spurious explanations instead. That, in turn, leads us to Hume's next charge against the false philosophy, which is that its theoretical posit has no *perceptible model* and therefore relies on an abuse of language.

> But as nature seems to have observ'd a kind of justice and compensation in every thing, she has not neglected philosophers more than the rest of the creation; but has reserv'd them a consolation amid all their disappointments and afflictions. This consolation principally consists in their invention of the words *faculty* and *occult quality*. For it being usual, after the frequent use of terms, which are really significant and intelligible, to omit the idea, which we wou'd express by them, and to preserve only the custom, by which we recal the idea at pleasure; so it naturally happens, that after the frequent use of terms, which are wholly insignificant and unintelligible, we fancy them to be on the same footing with the precedent, and to have a secret meaning, which we might discover by reflection. The resemblance of their appearance deceives the mind, as is usual, and makes us imagine a thorough resemblance and conformity. By this means these philosophers set themselves at ease, and arrive at

> last, by an illusion, at the same indifference, which the people attain by their stupidity, and true philosophers by their moderate scepticism. They need only say, that any phaenomenon, which puzzles them, arises from a faculty or an occult quality, and there is an end of all dispute and enquiry upon the matter.
>
> T 1.4.3.10; SBN 224

As we have seen, for Hume the ultimate aim of science is to satiate the desire for explanation. When this is accomplished legitimately, it is by explaining observed regularities via an appeal to the essence, nature, and powers of certain theoretical posits, and resting content when such explanations are no longer possible because there is no further perceptible model on which to base them. The false philosophy mimics this procedure by using the general terms "faculty" and "occult quality" to feign representing a substance that underlies the observed phenomena, where in fact there is no perceptible model on which the use of such terms is based, and therefore there is no actual content to these representations. "Faculty" and "occult quality" are words with no ideas associated with them (but with a promise that such ideas might discovered by reflection). So, again, the false philosophy purports to offer an explanation of the constancy and coherence of our perceptions, but in ignoring the explanations given by reason, it instead employs only the *form* of explanation without any of its content. It aims to construct a picture of the reality underlying the observed phenomena by employing certain general terms, but since these are not based on any perceptible model, its explanation is entirely spurious.

> Philosophers deny our resembling perceptions to be identically the same, and uninterrupted; and yet have so great a propensity to believe them such, that they arbitrarily invent a new set of perceptions, to which they attribute these qualities. I say, a new set of perceptions; For we may well suppose in general, but 'tis impossible for us distinctly to conceive, objects to be in their nature any thing but exactly the same with perceptions.
>
> T 1.4.2.56; SBN 218

It is *impossible* to represent anything that is not a perception. Doing so would require not just an analogical extension of experienced phenomena, but also the positing of a *via negativa*, something that is not specified to differ from perceptions in some determinate way (e.g., as being bigger, or of a different color, etc.), but rather something that is merely "different" from them. As we have seen, Hume considers difference to be no real relation at all, but rather only the negation of some determinate relation. Thus, the farthest that the false philosopher can go towards representing material substance is to invent a new set of perceptions that they misleadingly label "substance" or "occult quality".

Notice one more thing about this passage: in limiting the scope of our possible theoretical representative activity here, Hume once again intimates that the goal of such activity is not merely to describe and subsume observed phenomena under universal generalizations, but also to *distinctly conceive the nature of objects*. As it turns out, there is a rather severe limit on our ability to do this (that which we observed in 1.2.6)—the nature of objects cannot be any thing but exactly the same with perceptions—but the more general point remains that the *aim* of science, which one can successfully reach so long as it uses perceptions themselves as its model, is to explain observed regularities of experience by appealing to the nature of the mind itself.

We can now proceed to Hume's final critique of the false philosophy: that its purported explanatory force is in fact entirely parasitic on that of the system of the vulgar. That charge is itself derived from Hume's two-part observation that, "this philosophical hypothesis has no primary recommendation, either to reason or the imagination" (T 1.4.2.47; SBN 212). Hume's argument that the false philosophy has no recommendation to reason is that since reason concerns relations of cause and effect, and relations of cause and effect are based on past experience, and we have no past experience that allows us to represent material substance, there can be no reasoning that leads to the positing of material substance. That argument is a reiteration of the previous charge that we saw Hume level against the false philosophy. That the false philosophy also has no primary recommendation to the imagination is a conclusion that Hume finds more difficult to establish, but which he tentatively accepts nonetheless.[29] Hume's reasons for accepting that conclusion need not concern us here, though, as our focus will be on what Hume takes to follow from the combination of these conclusions.

> As to the second part of the proposition, *that the philosophical system acquires all its influence on the imagination from the vulgar one*; we may observe, that this is a natural and unavoidable consequence of the foregoing conclusion, *that it has no primary recommendation to reason or the imagination*. For as the philosophical system is found by experience to take hold of many minds, and in particular of all those, who reflect ever so little on this subject, it must derive all its authority from the vulgar system; since it has no original authority of its own.
>
> T 1.4.2.49; SBN 213

Hume's account of how the false philosophy usurps the explanatory force of the language of the vulgar is straightforward enough. The vulgar feel compelled to explain the apparent constancy and coherence of our perceptions, which leads them to suppose that these perceptions continue to exist when not perceived. While the explanation offered by the false philosophy purports to move beyond this explanation in positing certain "faculties" and "occult qualities," since these general terms have no content of their own, any explanatory power that they appear to have can only be that which is derived from the original supposition of the vulgar that *something* must continue to exist when not perceived.[30]

Finally, notice that Hume's claim here that since the false philosophy has no primary recommendation to reason or the imagination it has no authority of its own at least weakly implies that *should* a system have recommendation to reason or the imagination, it *would* also thereby have authority of its own, i.e., such a system would not depend for its acceptance on preserving the distinctions made in the language of the vulgar, but could instead replace these with its own picture of the world. These, of course, are precisely the grounds on which the true philosophy, the science of human nature, successfully supersedes the system of the vulgar.

Explaining "Body"

The vulgar system and that of the false philosophy are both attempts to explain the same phenomenon: the perceived constancy and coherence of our perceptions. Given what we already know about Hume's approach to the science of human nature, the failures of those explanatory theories appear to leave two options available to Hume: attempt to explain this *explicandum* via some third system of explanation, or, to resist the intemperate desire to attempt "explaining the ultimate principles of the soul" (T Intro.9; SBN xvii–xviii), and

> sit down contented; tho' we be perfectly satisfy'd in the main of our ignorance, and perceive that we can give no reason for our most general and most refin'd principles, beside our experience of their reality.
>
> T Intro.9; SBN xviii

As it turns out, the specific failures of the vulgar system and the false philosophy make these two options both unappealing and avoidable. It is unappealing to attempt a third explanation of the constancy and coherence of our perceptions because the two failed systems appear to exhaust the most plausible candidates for doing so. The vulgar attempt to posit a set of perceptions to do this explanatory work; the false philosophy posits a set of non-perceptions to do so. There are undoubtedly more subtle lines to pursue here, but broadly speaking, the prospects for explaining the behavior of our perceptions appears dim. If perceptions cannot explain this constancy and coherence, and non-perceptions cannot explain it, there seem to be few options remaining.

As for taking these regularities to themselves be explanatorily basic, as we have seen, this is never an easy option to take. The case of the constancy and coherence of our perceptions makes that even more difficult because of their apparent ubiquity and obviousness. Contrast the present case with that of the associations of ideas, which have a tempting specious explanation of their own that Hume eventually does manage to resist. The associations of ideas are "a gentle force, which commonly prevails," but which are also tempered by the freedom of the mind to join any two ideas,[31] and which must be posited by the science of human nature.[32] The constancy and coherence of our perceptions by contrast is, first of all, apparent to us at every moment without the least need of positing or discovery.

> There is scarce a moment of my life, wherein there is not a similar instance presented to me, and I have not occasion to suppose the continu'd existence of objects, in order to connect their past and present appearances, and given them such an union with each other, as I have found by experience to be suitable to their particular natures and circumstances.
>
> T 1.4.2.20; SBN 197

Second, the constancy and coherence is not only ubiquitous, but it also exerts a strong and almost subconscious demand for explanation owing to the high degree of force and vivacity of that its constituent resembling perceptions impart to each other.

> The mind falls so easily from the one perception [impressions] to the other [ideas], that it scarce perceives the change, but retains in the second a considerable share of the vivacity of the first. [. . .] Now this is exactly the present case. Our memory presents us with a vast number of instances of perceptions perfectly resembling each other, that return at different distances of time, and after considerable interruptions. This resemblance gives us a propension to consider these interrupted perceptions as the same; and also a propension to connect them by a continu'd existence, in order to justify this identity, and avoid the contradiction, in which the interrupted appearance of these perceptions seems necessarily to involve us.
>
> T 1.4.2.41–2; SBN 208

The constancy and coherence of our perceptions appears to be ubiquitous and obvious, and presents itself with a high degree of force and vivacity. Combined that with the attendant apparent contradictions that the actual interruptions in perception present, and the demand for an explanation of this constancy and coherence seems almost impossible to refuse. It is no wonder that Hume ends the *Treatise* by concluding that the return to specious and chimerical principles and former bad habits is both difficult to resist and entirely justified *practically*, if not theoretically. One could hardly expect even the scientist of human nature to resist the temptation to attempt an explanation of such phenomena.

Hume, however, sees a way out. The first step here is to recall that the purported constancy and coherence of our perceptions is entirely *specious*. It is a product of the imagination's tendency,

> when set into any train of thinking [. . .] to continue, even when its object fails it, and like a galley put in motion by the oars, carr[y] on its course without any new impulse.
>
> T 1.4.2.22; SBN 198

That is, recall that we saw earlier that while there is *some* constancy and coherence to our perceptions, because of the imagination's susceptibility to custom, habit, and education, the amount of constancy and coherence to be found amongst our perceptions is greatly exaggerated. That being so, it is not at all clear that the purported *explicandum* of the theories of the vulgar and false philosophy really exists at all, much less demands explanation. So, rather than rest content with taking the constancy and coherence of our perceptions as explanatorily *basic*, or attempt to find an explanation that is different from those of the vulgar and false philosophy, Hume can reject the claim that there is any such phenomenon in need of explanation. Since the constancy and coherence of our perceptions is not, in fact, an inductively-established universal regularity of experience, and is revealed to be not much of a regularity *at all*, there is no reason to seek explanations for it.

Still, there is more to Hume's account than merely that rejection. While the constancy and coherence of our perceptions may not be the proper object of scientific inquiry, the fact that our imagination inevitably *takes there to be* such a constancy and coherence, and we are therefore compelled to *believe* in the accounts of the vulgar or the false philosophy, is itself an empirically discovered regularity in need of explanation. As Hume notes at the start of 1.4.2,

> We may well ask, *What causes induce us to believe in the existence of body?* but 'tis in vain to ask, *Whether there be body or not?* That is a point, which we must take for granted in all our reasonings.
>
> T 1.4.2.1; SBN 187

Despite the failure of any attempt to explain the supposed constancy and coherence of our perceptions, we find ourselves compelled to believe that such regularities exist, and that there is something that can do this explanatory work. The resilience of *that belief* is itself a universal empirical regularity discovered among our perceptions that calls for its own explanation. Hume's explanation, of course, is that since many of our impressions resemble one another, the imagination forms a habit of moving quickly from one to the next, which in turn causes us to suppose that they are identical, which in its turn causes us to attribute their interruptions to ourselves rather than the distinct perceptions themselves. So, again, Hume relies on the theoretical apparatus that he has already established, including his positing of simple ideas and the faculty of the imagination, to explain an inductively-established universal regularity of perception. That regularity is not now the constancy and coherence of our perceptions, but our mistaken belief in that constancy and coherence and our attendant attempts to explain it.

After making his proposal, Hume goes on to note that,

> In order to justify this system, there are four things requisite. *First*, To explain the *principium individuationis*, or principle of identity. *Secondly*, Give a reason, why the resemblance of our broken and interrupted perceptions induces us to attribute an identity to them. *Thirdly*, Account for that propensity, which this illusion gives, to unite these broken appearances by a continu'd existence. *Fourthly* and lastly, Explain that force and vivacity of conception, which arises from the propensity.
>
> T 1.4.2.25; SBN 199–200

Once again, if all that Hume was doing were discovering the universal regularities that govern the human mind, rather than seeking to *explain* such regularities themselves, these tasks would not present themselves. Consider, for example, the third item on Hume's agenda: account for that propensity to unite these broken appearances by a continued existence. It is not enough that Hume discovers that we have a propensity to unite these appearances: he must also "account for" that propensity. Hume reiterates this distinction in the course of presenting this third item.

> We may begin with observing, that the difficulty in the present case is not concerning the matter of fact, or whether the mind forms such a conclusion concerning the continu'd existence of its perceptions, but only concerning the manner in which the conclusion is form'd, and principles from which it is deriv'd.
>
> T 1.4.2.38; SBN 206

Hume is not concerned to establish the matter of fact here, the empirical regularity, but rather his aim is to given an account of what the underlying mechanisms are (the manner and principles) that *explain* that fact. To do this, Hume first analyzes this explanatory demand into two complementary questions:

> Here then may arise two questions; *First*, How we can satisfy ourselves in supposing a perception to be absent from the mind without being annihilated. *Second*, after what manner we conceive an object to become present to the mind, without some new creation of a perception or image; and what we mean by this *seeing*, and *feeling*, and *perceiving*.
>
> T 1.4.2.39; SBN 207

Part of what an explanation of the fact that the mind forms the idea of a continued existence to explain the apparent constancy and coherence of our perceptions will include is an account of what precisely the underlying mechanisms are that produce this idea. Unsurprisingly, the answers that he gives to these questions is highly *theoretical* rather than merely descriptive.

> As to the first question; we may observe, that what we call a *mind*, is nothing but a heap or collection of different perceptions, united together by certain relations, and suppos'd, tho' falsely, to be endow'd with a perfect simplicity and identity. Now as every perception is distinguishable from another, and may be consider'd as separately existent; it evidently follows, that there is no absurdity in separating any particular perception from the mind; that is, in breaking off all its relations, with that connected mass of perceptions, which constitute a thinking being.
>
> T 1.4.2.39; SBN 207

Keep in mind here that what Hume is attempting to explain is the propensity of *the vulgar* to posit a continued existence. His explanation of that propensity is that the *since the mind is nothing but a bundle of perceptions*, the vulgar can formulate this idea by supposing that some of these perceptions exist outside of that bundle. Obviously, Hume does not take the vulgar themselves to explicitly cast their idea of body in this way. So Hume is not merely describing the observed behavior of the vulgar. Rather, he is giving a theoretical-explanatory account of it. We can observe the vulgar formulating such-and-such an idea of body: here is what must be going on in their minds when they do so, given the theoretical-explanatory system with which they are working.

Hume's answer to his second question is similarly theoretical.

> The same reasoning affords us an answer to the second question. If the name of *perception* renders not this separation from a mind absurd and contradictory, the name of *object*, standing for the very same thing, can never render their conjunction impossible.
>
> T 1.4.2.40; SBN 207

Here Hume combines the previous answer, which made an explicit and essential use of his own account of personal identity, with an implicit appeal to his account of general representation. Again, Hume moves beyond mere observation to pursue the truth about human nature that "must lie very deep and abstruse" (T Intro. 3; SBN xiv-xv). He explains the persistent tendency of the human mind to seek an explanation of the specious constancy and coherence of our perceptions by appealing to its underlying nature and behaviors: its being composed of unobservable simple ideas the associations of which give content its representations.

The entirety of Hume's treatment of the idea of body then reflects the realist methodology that we have discovered throughout Hume's pursuit of the science of human nature. What demands explanation are the inductively-established universal regularities that are discovered in experience. The imagination leads us to believe that we have discovered such a regularity in the constancy and coherence of our perceptions. This leads the vulgar to attempt to explain that regularity by forming a theoretical posit modeled on perceptions themselves. Their theory, however, which is a product of the unreliable imagination rather than reason, unsurprisingly ultimately contradicts the actual empirical evidence. The failure of the vulgar hypothesis prompts another attempt at explanation, that of the false philosophy, which posits body not as bearing a determinate difference from experienced perceptions, but as difference from them in all ways. As the scientific realist requires that theoretical posits by "founded on experience," the hypothesis of the false philosophy turns out to be nothing more than a mere manipulation of *words* with no underlying explanatory force (recommendation to reason or the imagination). What Hume's own investigation reveals is that the supposed constancy and coherence of our perceptions is not a discovery of reliable reason, but of the biased imagination, and is unsurprisingly entirely specious. It, therefore, does not call for explanation at all, although the regularity with which we fall prey to its illusions does. Hume explains *that* regularity in just the way that we would expect a scientific realist to do: via an appeal to the nature and behavior of the substance underlying the phenomenon, in this case the human mind itself.

Having accounted for the failures of the systems of the vulgar and false philosophy, and the success of the science of human nature in its turn, we have come a long way towards understanding Hume's methodology in conducting the science of human nature, and his understanding of the realist results of that methodology. Still, much of what I have argued thus far hangs on Hume's use of what we can call substantial explanation: explaining the inductively-established universal regularities of experience by appeal to the nature, powers, and essence of the substance underlying these manifest phenomena, and there are at least a few details of how that brand of explanation operates that need to be addressed, which will be the business of the next chapter.

Notes

1. "As this dilemma is an evident consequence of the principle, that every idea arises from a similar impression . . . " T 1.2.6.3; SBN 66
2. As Cummins, "Hume on the Idea of Existence," argues well, there are several difficulties attendant to understanding just what the nature of Hume's evidence for this claim is. For example, what Hume appears to need for this argument is that it is *impossible* to separate an idea of existence from any perception, but his evidence appears to be that he does not, *actually*, separate any two such ideas.

3. For example, Hume argues in 1.4.4 that since no external existence has secondary qualities, and secondary qualities are inseparable from primary qualities, no external existence has either primary or secondary qualities, and thus does not resemble any perception in any way. My thanks to Jonathan Cottrell for calling my attention to the relevance of that argument here.
4. My focus here will be on Strawson's treatment of "relative ideas," but similar interpretive lines are suggested in Wilson, "Is Hume a Sceptic with Regard to the Senses?"; Wright, *The Sceptical Realism of David Hume*; Craig, *The Mind of God and the Works of Man*; and Kail, *Projection and Realism*. More on Kail farther along.
5. Of course, even this much will be tricky business, as we saw in Chapter 1 that we can form no idea of a relation independently of its relata.
6. Strawson, *Secret Connexion*, 44–5.
7. Prominent early objections are raised in Winkler, "New Hume"; Blackburn, "Hume and Thick Connections"; Flage, "Relative Ideas Re-Viewed"; and Millican, "Hume, Causal Realism, and Causal Science."
8. Strawson, *Secret Connexion*, 46–7.
9. For example, in his Chapter 12 on meaning.
10. Rosenberg, *Beyond Formalism* is an unfortunately overlooked but excellent extended critique of Kripke's theory of reference to which what follows owes a great deal.
11. Kripke, "Speaker's Reference and Semantic Reference," 264.
12. "the reference of the idea to an object being an extraneous denomination, of which in itself it bears no mark or character" (T 1.1.7.6; SBN 20).
13. As Boehm points out, this implies that if it were to turn out that that unknown causes of our impressions were, e.g., "the creative power of the mind" or "the author of our being" (T 1.3.5.2; SBN 84), then such a relative idea would not, in fact, refer to external objects at all. Boehm, "Concept of Body in Hume's *Treatise*," 208. Along the same lines, it is worth noting that Strawson's take on relative ideas also implies that they are not, in fact, rigid designators. Strawson himself acknowledges as much. Strawson, *Secret Connexion*, 122. See also, Flage, "Hume's Relative Ideas."

    In discussing this same point, Kail comes ever so close to hitting upon precisely what is needed to correct Strawson's account: the notion of using a perceptible model that is specified to resemble and differ from a theoretical posit in determinate ways.

    > Rather than trade on the supposition of 'unknown somethings', the realist should instead simply point out that the supposed external objects *resemble* perceptions. That prevents the objection that supposed 'causes of perception' is too under-determined to bear any weight. The difference between perceptions and external objects is captured in two ways. First, such objects are like perceptions except that they have 'different relations, connexions and durations': that is, they continue to exist unperceived and are distinct. [. . .] Secondly, the notion of a relative idea allows the thinker to think that there may be facts about such objects over and above those represented in perception by resemblance.
    >
    > Kail, *Projection and Realism*, 60.

    Kail is absolutely right to notice that if a theoretical posit is to do the explanatory work that Strawson assigns to it, it will need to resemble some perceptible model in some determinately specified ways. The problem with Kail's suggestion is that he allows for the *differences* between the model and the posit to be left entirely *indeterminate*. The "different relations, connexions, and durations" alluded to cannot, in fact, be given any determinate specification, and the "facts over and above those represented in perception by resemblance" will be ones that we cannot, *ex hypothesi*, ever so much as even possibly represent. The problem here is not that the use of a model leaves open certain questions for further investigation. That is an essential part of the role of scientific models. Rather, the point is that the theoretical posit cannot be given *any* determinate content at all precisely because it is supposed to resemble perceptions, but be different from them only in ways that are entirely *inarticulable*.

Worth noting is that Kail too eventually makes an appeal to Kripke in defending his view. Like Kripke, Kail finds himself needing to account for what appears to be our capacity to conceive of different effects following from like causes despite the fact that he holds that this is in fact impossible. (Thus, Kail also needs to explain away the apparent close tie that Hume draws between conceivability and possibility.) Following Kripke, Kail holds that we are not conceiving the *kinds* which underwrite laws being followed by different effects (a metaphysical impossibility), but "putting ourselves in a position whereby something different, but with the same sensible qualities, is followed by such and such effects" (Kail, *Projection and Realism*, 98). Of course, to separate out the sensible qualities of an object from its underlying nature, Kail employs the Bare Thought, his improvement over Strawson's take on relative ideas. See below.

14. Garrett, *Cognition and Commitment*, 106.
15. Strawson, *Secret Connexion*, 48.
16. A similar point as this final one was first made in Winkler, "New Hume," and revisited in Garrett, *Cognition and Commitment*.
17. Strawson, *Secret Connexion*, 123.
18. A similar point speaks against Kail's attempt to improve upon Strawson. Kail concedes that Hume holds that relative ideas are "meaningless," but nonetheless contends that we can have knowledge of what it would be to encounter their objects.

> The weaker interpretation of meaninglessness has it that the term is meaningless because the object is not a possible object of acquaintance but nonetheless grants that we can form some sufficiently rich notion to ensure we are thinking about the object that is not a possible object of acquaintance. [. . .] This thin notion—which is not an idea—expresses itself in the knowledge of what it would be to be acquainted with the relevant object.
>
> (Kail, *Projection and Realism*, 34–5)

Kail casts the objects of such a meaningless term as being not even so much as the *possible* objection of acquaintance, but also holds that we can *know* what it would be to be acquainted with such objects. Put aside for the moment that this appears to imply that we can know what it would be like for something impossible to happen; Kail is still left with a more parochial problem. To *know* such a thing, we would have to be able to believe it. For Hume, however, a belief is a lively *idea*. Kail makes a point of noting that whatever such representations are, they are not ideas. Thus, such a representation, or non-idea, cannot be, or become, knowledge. Nor can we have "knowledge of what it would be to be acquainted with the relevant object," because such knowledge itself would have to be an idea, presumably a complex one with the non-idea representation as one part and the object as another, which idea is impossible for the same reasons.

Furthermore, when Kail turns his attention specifically to this contrast between supposition and belief, his position does not improve.

> 'Supposing' here is not understood to have the force of justified belief, but something far weaker. First, it is not *believing*: his official account of the state of *belief* requires that any such belief in causal power to involve such an idea to combine to deliver the conclusion that no such belief is available to Hume. Any considered realist position will insist on attitudes like supposition, assumption or taking for granted, attitudes that independently of the details of Hume's view of belief have different functional roles from belief.
>
> (Kail, *Projection and Realism*, 123)

As we have seen, it is important for the New Humean to deny that we can *believe* in external existence, casual powers, etc. because beliefs are ideas, and have descriptive content, whereas whatever the vehicle it is that is meant to represent, or refer to, such things is not an idea and does not have such content. So, Kail is surely right that his brand of realism must countenance doxastic attitudes other than belief. His proposal for *how* these attitudes differ from belief, though, appears to miss the mark entirely.

> Beliefs tend to govern one's behavior in a way that assumptions do not: one may be more prepared to defend the former, or more likely to retain beliefs in the face of challenges than assumptions or allow them to figure in practical as opposed to speculative inference.
>
> (Kail, *Projection and Realism*, 123)

The main problem with this way of understanding the difference between belief and supposition is that casts that difference as consisting entirely of a difference in their *extrinsic* properties, but what the New Humean needs is precisely an interpretation of supposition and belief that accounts for their *intrinsic* differences (most importantly, that suppositions are not *images*, and therefore have no descriptive content).

Another problem with it is that Kail repeatedly claims that we *do* believe in such things—in his chapter on necessary connection alone he does so on pages 116, 117, 122, and 124—and even proposes an account of precisely how he takes such beliefs to be justified! He argues that while reason does not provide an adequate justification for such beliefs, "it is very plausible to think that Hume does not recommend the unrestricted deliverances of reason as authoritative on belief" (Kail, *Projection and Realism*, 124), and that such beliefs are also supported by practical considerations. All of that is to say, however, that Kail's distinction between supposition and belief also cannot do the work that he and other New Humeans need it to do.

19. One last point on Kail's counterpart to Strawson's relative ideas: the Bare Thought.

> We can specify uniquely that which we cannot understand (causal power) by saying that it is that feature that, were we acquainted with it, would yield *a priori* inference and render it inconceivable that the cause not be followed by its effect. This is the Bare Thought.
>
> (Kail, *Projection and Realism*, 84)

Kail does an excellent job rebuffing potential objections to this proposal, and I think it worth following him through at least one dialectical thread. Here is an initial objection to the Bare Thought. If, as Kail writes, "we have absolutely no idea of is what kind of thing it is that could, through our detection of it, yield those consequences" (Kail, *Projection and Realism*, 84), then might we not think that the notion of *a priori* knowledge of a causal relation is like that of a square circle? It *seems* like the phrase "square circle" describes something by combining two internally coherent ideas, but that is an illusion. In fact, it can be shown that what is knowable *a priori* is such only because it is a relation of ideas, and that causal relations are not, and could not be, such relations.

> the notion of a feature, acquaintance with which would close down our imaginative capacities, is itself impossible, since all possible objects of *acquaintance*—impressions—are phenomenally distinct, loose and separable items. So we cannot be acquainted with anything that would allow us to make such *a priori* inferences and close down what we presently find conceivable. Our candidate for necessity is incoherent, and we are left with no other.
>
> (Kail, *Projection and Realism*, 99)

Kail has a ready reply. The objection fails because Hume holds that we can make sense of the idea of powers of representation other than our own, e.g., "the rationalist faculty of the intellect, whereby non-sensory representations provide content for thought" (101). Putting it this way brings out exactly why Hume would find such an alternative incoherent. Hume not only rejects the suggestion that *we* have such a faculty, but also that any such faculty is so much as *possible*. Representation, for Hume, *just is* copying, etc. What *it is* for one thing to represent another is to be a copy of it. So, the notion of a representation that is nonimagistic, and therefore not a copy, *is* incoherent. In Landy, "Hume's Theory of Mental Representation," I detail how Hume can arrive at the Representational Copy Principle via an *analysis* of the meaning of "representation".

20. Strawson, *Secret Connexion*, 18.
21. He also holds that what explains these regularities must be some underlying *necessary* connections. In the next chapter, we will see that this too is a thesis that Hume rejects.
22. As I understand what Hume is after here, it is that the imagination is responsible for both misleading the vulgar with respect to the *explicanda* at hand—that our perceptions exhibit a high degree of constancy and coherence—and its *explanans*—that this supposed constancy and coherence of our perceptions is explained by the distinct and continued existence of some other perceptions. Hsueh Qu points out that this second claim amounts to the imagination's attempting to explain some phenomenon, which I cast in the previous chapter as the defining proper function of reason. Notice that while it is the defining proper function of reason to explain, this can also be a task that is *improperly* given over to the imagination, just as one can use a clock as a doorstop, even if it is neither designed nor well suited for that purpose.
23. Here is another passage in which Hume contrasts the reliable deliverances of reason (and philosophical principles) with the unreliable ones of imagination, and again specifically with respect to the idea of body.

> We may also observe in this instance of sounds and colours, that we can attribute a distinct and continu'd existence to objects without ever consulting REASON, or weighing our opinions by any philosophical principles. [. . .] Accordingly we find, that all the conclusions, which the vulgar form on

> this head, are directly contrary to those, which are confirm'd by philosophy. For philosophy informs us, that every thing, which appears to the mind, is nothing but a perception, and is interrupted, and dependent upon the mind; whereas the vulgar confound perceptions and objects, and attribute a distinct continu'd existence to the very things they feel or see.
>
> T 1.4.2.14; SBN 193

Philosophy and reason *confirm* and *inform* us that there are no distinct continued existences. Those are both success verbs, and Hume fairly clearly endorses the conclusions of philosophy in this regard, as they are the very ones that he himself presents in 1.2.6 and 1.4.2. Furthermore, the conclusion of the paragraph in which this passage appears is that the idea of body that the vulgar employ "must be entirely owing to the IMAGINATION" (T 1.4.2.14; SBN 193). That is, being a product of the imagination, as opposed to senses or reason, makes the idea of body, while ubiquitous, also *false*.

24. So no third dimension after all.
25. Butler presents an argument that would count against understanding the vulgar fiction of continued existence as requiring the kind of theoretical-explanatory activity that I here describe.

> Hume criticizes competing accounts in "that they suppose such a subtility and refinement of thought, as not only exceeds the capacity of mere animals, but even of children and the common people in our own species" (T 1.3.16.3; SBN 177). Thus, if human causal inference about bodies presupposes belief in body, so does animal causal inference. Therefore, Hume's psychological account of belief in body must draw on resources accessible to animals as well as humans. This places constraints on a suitable interpretation of his account of the belief in body and a fortiori the "fiction" of continued existence. Specifically, the account cannot depend on language, abstraction, or complicated reflective reasoning.
>
> Butler, "Hume on Believing the Vulgar Fiction of Continued Existence," 237.

Since I hold that theoretical representation essentially depends on the use of general terms and perceptible models, I am thereby committed to this fiction's depending on language, abstraction, and complicated reflective reasoning. As I argued in the previous chapter, however, I take these to be compatible with Hume's claim that our reason is of a piece with that of animals because I understand Hume as holding that some animals also engage in something like an impoverished version of theoretical-explanatory reasoning. While dogs operate on mere custom, birds engage in a distinct kind of reasoning like that of the chemist. Thus, I take Hume's admonishment of those who, "suppose such a subtility and refinement of thought, as not only exceeds the capacity of mere animals, but even of children and the common people in our species," to address not the supposition that the vulgar fiction relies on theoretical-explanatory reason at all, but rather that it employs a particularly complex and sophisticated theoretical apparatus. As we have seen, "common language has seldom made any very nice distinctions." It is not that vulgar do not employ theoretical-explanatory posits, but rather that they are *clumsy* when they do so.

That said, I also think there is independent reason to reject the account of the vulgar fiction that Butler offers.

> The mind supposes that a single object is conjoined with a duration. The duration is supplied by the successive perceptions that I experience,

including the table impressions and my blinking. But rather than multiplying the table perceptions, the imagination takes one single impression of sense or memory and treats it as steadfast.

Butler, "Hume on Believing the Vulgar Fiction of Continued Existence," 241.

Butler's idea is that rather than take the table to consist of the temporal succession of qualitatively identical table impressions, we take it to consist of a single "steadfast" table impression. What puzzles me here is what this can mean. Simple impressions are temporally minimal, so insofar as the table impression is extended in time, it is complex. Butler cites Baxter's discussion of steadfast objects here, and Baxter is explicit on this point.

> A succession is a number of distinct things, one after the other. In contrast are what Hume calls "stedfast" and "unchangeable" objects. These are single objects that coexist with successions but which are not successions themselves. Thus they lack duration despite coexisting with successions that, as such, have duration. A mantelpiece would be an example of a steadfast object coexisting with the flickering succession of flames below. It would coexist likewise with the less predictable succession of thoughts in the mind of the homeowner relaxing alongside.
>
> T 1.2.3.6–11; SBN 34–7

Again, as I understand Hume, he takes the steadfast mantelpiece to consist of a temporal succession of simple impressions, and so to be a complex, not a simple idea. Note that that thesis does not require adopting the notion of the simple ideas as theoretical posits with which I have been working. Even on a more standard account, since we can distinguish the temporal parts of the mantelpiece, e.g., the temporal part of it that coincided with the flames leaping up the temporal part of it that coincided with the fire dying down, etc., those parts are separable and therefore distinct. Thus, the vulgar fiction cannot be as Butler describes it to apply, "to a single unchanging impression the idea of duration, which, properly speaking, applies only to successions." The steadfast objection *is* a succession, so this cannot be why the idea of duration does apply to it. And, of course, adopting the thesis that simple ideas are theoretical posits makes this case, in one sense, even easier to handle: even if we cannot *observe* the simple perceptions that compose the mantle, we can still *know* that it is composed of them.

26. "We suppose external objects to resemble internal perceptions. [. . .] We never can conceive any thing but perceptions, and therefore must make every thing resemble them" (T 1.4.2.54; SBN 216).
27. My thanks to Jonathan Cottrell for raising both facets of this line of objection.
28. Loeb argues that constancy ought to be treated as a special case of coherence. Loeb, *Stability and Justification*, 207–14.
29. T.1.4.2.48; SBN 213.
30. In defending a very different "Sellarsian" version of Hume, Wilson offers a different take on this passage. He takes "the philosophical system" to refer not just to the "false philosophy," but to philosophical systems in general (as opposed to the system of the vulgar) including Hume's own. Thus, as Wilson understands this passage, when Hume writes that the philosophical system has no *primary* recommendation to reason, this leaves open the possibility that it can still have a *secondary* recommendation to reason.

> The system of the philosophers distinguishes perceptions, i.e., impressions, that exist only when perceived, and objects that have a continued existence.

> Now, Hume points out, since the mind has no acquaintance with the latter objects, custom can have no direct force in leading us to a causal inference to those objects. The system therefore has no primary recommendation to causal reason (212). It does not follow that it has no appeal to reason; for, it could still have a secondary recommendation to reason. The relevant habits of causal inference must be acquired elsewhere, and this elsewhere can only be the system of the vulgar: thus, "the philosophical system acquires all its influence on the imagination from the vulgar one" (213). The point is that reasoning within the system of the vulgar will lead one outside that system to another, contrary, system.
>
> Wilson, "Is Hume a Sceptic with Regard to the Senses?," 56

As we have seen, Wilson is exactly right that the reasoning of the vulgar "will lead one outside that system to another, contrary system," i.e., that such reasoning leads to a contradiction that the false philosophy is meant to correct. He is led astray, however, by rejecting the distinction at work here between the false philosophy and the science of human nature. The false philosophy posits theoretical entities (objects) that in no way resemble perceptions. Thus, the representations of such entities are devoid of all content. Wilson holds—with other New Humeans—that such a posit is possible and legitimate because it is made via a "relative idea" (Wilson, "Is Hume a Sceptic with Regard to the Senses?," 57), which itself has no content, but nonetheless refers to that-which-is-the-cause-of-my-impressions. We have already seen why this gambit cannot work.

31. T 1.1.4.1; SBN 10.
32. Since the associations of ideas hold primarily of simple ideas, which are themselves theoretical posits. (T 1.1.4.1; SBN 10.) The dialectic in which the associations of ideas are posited is similar to that in which the simple-complex distinction is: both begin with Hume proposing that a great deal about human experience could be explained upon the supposition of the truth of the hypothesis. "As all simple ideas may be separated by the imagination, and may be united again in what form it pleases, nothing wou'd be more unaccountable than the operations of that faculty, were it not guided by some universal principles, which render it, in some measure, uniform" (T 1.1.4.1; SBN 10). It is only after demonstrating the explanatory demand for such principles that he then goes on to present the evidence from experience that confirms their operations.

5 Necessary Connection and Substantial Explanation

The purpose of the present chapter is to confront the apparent clash between two central theses of my interpretation. On the one hand, I have argued that all theoretical representation must proceed via the use of a perceptible model. Theoretical-explanatory posits are represented by specifying the determinate ways that such entities both resemble and differ from some previously-experienced phenomena. Hume repeatedly rejects as specious and chimerical those posits of his predecessors that cannot be so represented. In the previous chapter we saw one such casualty: the explanations of the apparent constancy and coherence of our perceptions given by the vulgar and the false philosopher. In the current chapter, we will likewise see Hume reject the notion of necessary connection on the grounds that there is no experience that can serve as the model for such an idea. Roughly, using the schema from previous chapters to represent two items as necessarily connected would require that the ideas of those items themselves be necessarily connected, i.e.,

> 'x' necessarily connected to 'y' represents x necessarily connected to y.

As Hume observes, no distinct ideas are necessarily connected to one another, so this schema can never be instantiated, so we cannot represent necessary connection. Importantly, this implies that to represent two items as *causally* related is not to represent them as being necessarily connected, and that is a thesis that might appear to come into conflict with one of the core principles of the interpretation of Hume's science of human nature that I have been defending, namely, that theoretical science proceeds by explaining the observed regularities of experience by appealing to the nature of the *substance* underlying such phenomena.

That thesis appears to be in tension with the previous one because it is plausible that such substantial explanations proceed by casting the observed regularities as *necessarily following from* the nature of the underlying substance. For example, when we explain the fact that gases obey the Ideal Gas Law by casting them as a collection of molecules themselves obeying familiar mechanical laws, it is plausible to suppose that a gas's being a collection of molecules, etc., *makes it necessary* that it obeys the Ideal Gas Law. That is, it *follows from* a gas being a collection of molecules, and the familiar mechanical laws, that gases *will* obey the Ideal Gas Law. Of course, were that the case, it would also follow that a gas being in a certain state at a certain time would be necessarily connected to its being in a different state at a different time. I believe that it is something like that thesis that underlies the following claim from Strawson.

> As it is, Hume never really questions the idea that there is something like causal power in reality, i.e. something about reality in virtue of which reality is regular in the way it is. [. . .] Hume believed in the existence of something like causal power or 'natural necessity'.[1]

I agree with Strawson that Hume holds that there is something about reality in virtue of which reality is regular in the way it is. Hume endorses and employs substantial explanation as the proper methodology for the science of human nature. *Contra* Strawson, though, I will argue that Hume's use of this form of explanation cannot require belief in, or even representation of, causal power or natural necessity. As noted, though, there does appear to be a tension between those two interpretive theses because it does not seem implausible to think that casting observable phenomena as being the manifestation of the nature of the substance underlying them makes them necessary.

To bring this tension out in another way, one might pursue the following line of objection. If it is the inductively-established universal regularities of experience that demand explanation, and this explanation cannot make appeal to the necessary connections between objects in accounting for such regularities, then it would seem that what will do this explaining will be the regularities governing the underlying substance. So, it is regularities all the way down. If that is right, though, then the regularities of substance appealed to as the *explanans* of the regularities of experience will themselves stand in need of explanation, etc. Thus, we set forth on a path that can only lead to an infinite regress, a circular explanation, or an arbitrary stopping point. Thus, it seems that one must forego either the commitment to substantial explanation or the thesis that we cannot represent necessary connections.

As I said, the purpose of the current chapter is to resolve this tension. The first step of this resolution will take the form of showing that it is only by presupposing that manifest natural laws require necessary connections between distinct items that one would be tempted to think that substantial explanations require the same, and since Hume rejects the former thesis, we should likewise understand him as rejecting the latter. That is, in representing the substance underlying manifest phenomena, we represent a certain causal structure, but the notion of causation here can be understood along thoroughly Humean lines, as involving no idea of necessary connection.

Of course, some philosophers take the very purpose of substantial explanation to be to represent the necessary connections among distinct existences, so if Hume rejects that thesis, then one can legitimately ask what he takes its aim to be. I will suggest that for Hume the proximate purpose of a substantial explanation is to render the manifest phenomena *expected*. Merely observing some regularity produces a low level of expectation the support for which extends only as far as our inductive inferences can carry it. By contrast, substantial explanation, by explaining the observed regularities via the *nature* of the underlying substance produces a more fixed and stable expectation that also supports a more robust epistemic reliability (which furthers the distal end of such explanations, accurately reflecting the true ontology of the mind). Even if the regularities governing the underlying substance will themselves demand explanation at some point, an explanation that locates these regularities in the nature of that substance temporarily forestalls that explanatory demand, and thus provides relief of the desire for explanation while also locating the source of the regularity in a way that makes for more predictive inferences.

That interpretive thesis reveals the means to address the second way of framing this tension. The concern is that if the *explicanda* of substantial explanation are the universal regularities discovered in experience, and what explains these are the regularities posited among an underlying substance, then it appears as though a vicious regress of explanation ensues. I will argue that Hume is well aware of this threat and addresses it by drawing a distinction between arguments from analogy, which make faux-explanatory appeals to "more of the same," and substantial explanations, the ontological implications of which carry real explanatory power and provide a non-arbitrary stopping point for explanatory demands. That is, a merely observed regularity demands an explanation because it is not taken to be an intrinsic feature of any particular kind of object. Nothing about being a gas makes it expected that a gas will obey the Ideal Gas Law, thus its doing so is *surprising* and puzzling. By contrast, when we explain these observed regularities of experience by appealing to the nature of the underlying substance, we cast the manifest phenomena as being of a kind the nature of which is to exhibit just such regularities. Whereas there is no reason to think that gases will obey the Ideal Gas Law other than our observing them to have done so, in reconceiving a gas as a collection of molecules subject to familiar mechanical laws, we take it to be the *kind* of object that behaves in the way that it in fact is observed to behave. Importantly, we attribute this behavior to the *nature* of the object so conceived. The key here is in recognizing the difference in the relative *stability* of the expectations formed in light of the different ways of locating the regularities. A merely observed regularity is only as stable as its inductive base; a regularity that is attributable to the nature of some substance, that is common to all objects of that kind, produce a more stable and reliable set of expectations. As Hume puts it, this particular kind of causal reasoning represents a "reality" that "admits not of the least change."

Before turning to the details of this resolution, though, I will first present my understanding of Hume on the impossibility of the idea of necessary connection. Following that, I explicate the nature of substantial explanation *sans* the idea of necessary connection. In the final section, I return to the framing puzzle with which this study long ago began, the puzzle concerning Hume's theory of general representation discussed in Chapter 1, to bring together the many theses that I will by then have defended into a single cohesive solution to that puzzle.

The Idea of Necessary Connection

Before turning to Hume's *arguments* concerning necessary connection, I want to spend some time focusing on his *conclusion*. As in previous cases, I hold that Hume's conclusion is first and foremost neither epistemic nor metaphysical, but rather it is a thesis about our representative capacities. We have no idea the representational content of which is necessary connection. As Hume himself indicates, the thesis that we have no idea of a necessary connection between distinct existences is a difficult one to articulate since our not having that idea implies that we cannot actually understand what idea it is that Hume concludes we do not have. As Hume sees it, what is needed in such cases is an account of how it is that we *take* ourselves to have such an idea, and in what that taking actually consists. As is usual in such cases, for reasons that we have seen in previous chapters, Hume focuses on the general terms that we use in describing the objects of such ideas. Here is the conclusion that he draws about such terms.

> Thus upon the whole we may infer, that when we talk of any being, whether of a superior or inferior nature, as endow'd with a power or force, proportion'd to any effect; when we speak of a necessary connection betwixt objects, and suppose that this connexion depends upon an efficacy or energy, with which any of these objects are endow'd; in all of these expressions, *so apply'd*, we have really no distinct meaning, and make use only of common words, without any clear and determinate ideas.
>
> T 1.3.14.14; SBN 162

Hume's conclusion here, the reasons for which we will see in a moment, is that when we *speak* of necessary connection, these *expressions so applied* have no distinct meaning, and make use only of common words without any clear and determinate ideas. His conclusion is one about the meaning of the words "necessary connection" (and its synonyms), and that conclusion is that as those words are used by the vulgar the ideas that are associated with them are a hodgepodge of confusion and indeterminacy. Hume's conclusion is that the vulgar's words "necessary connection" do not represent anything.

Of course, Hume complicates matters a moment later by announcing that whereas the words "necessary connection" do not have any meaning *as they are employed by the vulgar*, it is possible to recover their "true meaning"—the meaning that they are given by the "true philosophy"—by investigating what ideas would underlie a proper use of them.

> But as 'tis more probable, that these expressions do here lose their true meaning by being *wrong apply'd*, than that they never have any meaning; 'twill be proper to bestow another consideration on this subject, to see if possibly we can discover the nature and origin of those ideas, we annex to them.
>
> T 1.3.14.14; SBN 162

As Hume sees it, the importance of the words "necessary connection" derives from their association with the idea of causation, which we have seen is an idea fundamental to the scientific enterprise as he understands it. Thus, a recovery of the true meaning of "necessary connection" comes in the form of a proper definition of causation, which Hume goes on to give. Here, as elsewhere, Hume takes a bifurcated approach to the controversial terms at hand: he rejects the meaning and use that the vulgar give to those terms, while also appropriating the very same terms for use in the science of human nature. Here is how he puts his ambivalent conclusion farther along.

> Either we have no idea of necessity, or necessity is nothing but that determination of the thought to pass from causes to effects and from effects to causes, according to their experienc'd union.
>
> T 1.3.14.22; SBN 166

Hume here puts his conclusion in terms of a disjunction: either we have no idea of necessity *or* necessity is nothing but that determination, etc. An important question about this disjunction is what would make either of its disjuncts true and the other false. That is, what is the difference between our having no idea of necessity and our idea of necessity being nothing but a determination of thought? As the previous passage indicates, what makes this difference is the language in which one considers the idea of necessary connection. The vulgar take themselves to have an idea of necessity and so also take themselves to be able to coherently formulate the thesis that we have no idea of necessary connection, even if they are wrong that they can do so and take this thesis to be false. By contrast, the scientist of human nature, or the true philosopher, denies that there is any sense to be made of such an idea: in his adopted idiom "necessity" is nothing but that determination of the thought, etc. As we have seen repeatedly over the past few chapters, the language of the scientist of human nature replaces the language of the vulgar, and thereby recasts the vulgar's ontological commitments (in this case to powers or energies) as modes of the entities posited by that science (the determination of thought). As the scientist sees it, what the vulgar confusedly *take* to be features of "external existence" are actually features of the mind itself, and the ontological commitments of each group are reflected in the language that each employs, and in the replacement of the "common terms" with those of the true philosophy. So, Hume's disjunction is verified: speaking in the language of the vulgar, the scientist concludes that we have no idea of necessary connection, but speaking more properly, in the language of science, the scientist concludes that our general term "necessity" refers only to that determination of thought, etc.

Of course, Hume might not intend this as real disjunction, but rather as a way of saying that having no idea of necessity is the same as necessity's being that determination of thought. For example, perhaps a more perspicuous way of conveying Hume meaning would be something like, "Either we have no idea of necessity, or *to put it another way* necessity is nothing but that determination of thought," etc. Again, though, this would fit with the picture of language replacement with which we have been working. Having no idea of necessity is equivalent to our idea of necessity being a determination of thought precisely because to cast "necessity" as a determination of thought, as the scientist of human nature does, is to abandon the vulgar's confused and indeterminate use of "necessity" and so render their use of that term null and void. Our having no idea of necessity, and our idea of necessity's being an idea of a determination of thought, are equivalent just in case we have no idea of necessity as the vulgar use that term, but our idea of necessity as the scientist of human nature understands it represents a determination of the mind.

Recall that in discussing the idea of body in the previous chapter we had occasion to note Hume's interesting thesis in the following passage.

> The persons, who entertain this opinion concerning the identity of our resembling perceptions, are in in general all the unthinking and unphilosophical part of mankind, (that is, all of us, at one time or other) and consequently such as suppose their perceptions to be their only objects, and never think of a double existence internal and external, *representing and represented.*
>
> T 1.4.2.36; SBN 204, emphasis mine

In attempting to formulate a theoretical posit to explain the apparent constancy and coherence of our perceptions, the false philosopher casts the vulgar as confusing their represent*ings* (perceptions) with that which is thereby represent*ed* (material objects). While Hume rejects the false philosophers' positing of material beings, he does appear to accept part of their diagnosis of the vulgar's misstep. The vulgar to do not distinguish between the apparent constancy and coherence of their represent*ings* (produced by the inertial force of the imagination) and the apparent constancy and coherence of that which is thereby represent*ed* (the actual disjointedness of experience itself). It is easy enough to discern Hume making a similar move here, especially in light of the vulgar's confused conception of the nature of body. The vulgar mistake a feature of their represent*ing*—a determination of thought—for a feature of what they take to be thereby represent*ed*—a real connection between distinct (external) existences. In so doing, they talk in a way that belies the fact that the ideas that are, in fact, associated with the terms that they use are simply a confused mishmash of representings, representeds, the idea of body, the purported idea of external existence, the determination of thought, etc.

Here is one final passage that expresses this same view again, this time in the context of Hume's reply to an imagined interlocutor who insists on both parts of the vulgar's mistaken view of necessity: first that the term "necessity" has the meaning that the vulgar take it to have, and second that it refers to a power or efficacy in external existences.

> I can only reply to all these arguments, that the case is here much the same, as if a blind man shou'd pretend to find a great many absurdities in the supposition, that the colour of scarlet is not the same with the sound of a trumpet, nor light the same with solidity. If *we really have no idea of* a power or efficacy in any object, or of any real connexion betwixt causes and effects, 'twill be of little purpose to prove, that an efficacy is necessary to all operations. *We do not understand our own meaning in talking so, but ignorantly confound ideas, which are entirely distinct from each other.* I am, indeed, ready to allow, that there may be several qualities both in material and immaterial objects, with which we are utterly unacquainted; and if we please to call these *power* or *efficacy*, 'twill be of little consequence to the world. But when, instead of meaning these unknown qualities, we make the terms of power and efficacy signify something, of which we have a clear idea, and which is incompatible with those objects, to which we apply it, obscurity and error begin then to take place, and we are led astray by a false philosophy. *This is the case, when we transfer the determination of the thought to external objects, and suppose any real intelligible connexion betwixt them; that being a quality, which can only belong to the mind that considers them.*
>
> T 1.3.14.27; SBN 168

Since this interlocutor does not object to the soundness of Hume's arguments, but only to the purported absurdity of its conclusion, Hume responds appropriately by pointing out if the conclusion is accepted, then such objections can only be made from ignorance. It is not Hume's response that is of interest, though, but rather what he takes his own conclusion to be. "We really have no idea of a power or efficacy in any object, or of any real connexion betwixt causes and effects." Once again, Hume explicitly puts his conclusion in representational terms. His argument is intended to show that we *have no idea* the content of which is a necessary connection. Insofar as we take ourselves to have such an idea, "we do not understand our own meaning in talking so, but ignorantly confound ideas, which are entirely distinct from each other." The vulgar use words in a way that leads them to believe that what they are representing is a real connection between distinct existences, but really they merely confuse features of their represent*ings* with features of that which they attempt to represent: "we transfer the determination of thought to external objects [. . .] which can only belong to the mind that considers them." What Hume describes here is exactly the confusion that he earlier ascribes to the vulgar: confusing a feature of a represent*ing* with a feature of that which is represent*ed*, and that in turn implies that the language of the vulgar does not in fact represent necessary connection at all—the determination of thought is a feature of the represent*ing*, not the represent*ed*—but instead merely takes itself to do so by failing to reflect the relevant distinctions in its use of general terms.[2]

Of course, the interpretation of Hume as holding that we have no idea of necessary connection (as the vulgar attempt to use that term), or that our idea of necessary connection (as the scientist of human nature uses that term) is just an idea of the determination of the mind is not new, and in fact has been around quite long enough to receive its share of criticism. Not much of this criticism, however, is particularly persuasive.[3] Consider first, Broughton's objection to what she calls the Psychological Interpretation.

> The objection to the view is simple. People just *aren't* talking about their feelings when they talk about necessary connections. Suppose there are and even can be no necessary connections between objects. It is perhaps mysterious, then, what we mean by 'necessary connection.' Still, we know *enough* about what we mean to know that we are not making a claim about our feelings when, for example, we deny that there are necessary connections between objects.[4]

Broughton's objection, I would suggest, is not as simple or straightforward as she takes it to be. Consider what evidence Broughton seems to present for her claim that "people just aren't talking about their feelings when they talk about necessary connection." That evidence appears to be introspective: "we know enough about what we mean to know that we are not making a claim about our feelings." If my interpretation of Hume is correct, then such straightforward introspective evidence ought not to be treated as sufficient for making this point. Because the vulgar *lack the distinction* between representing and represented, when they introspect, they too find nothing wrong with the way they use the words "necessary connection". They take themselves to use those words entirely consistently, but do so only because they lack the distinction that is necessary for making clear where it is that their thinking is confused.

Furthermore, the vulgar would likewise agree with Broughton's claim to know that their attributions of necessary connection are not aimed at feelings, but rather at real connections between objects. Of course, as we have seen, insofar as the vulgar have a coherent idea of body, it is the idea of perceptions themselves, and so to this extent again the vulgar would be wrong about the content of their own ideas, and therefore wrong about the meaning of the term 'necessary connection'. My claim is not that the vulgar *take themselves* to be talking about a determination of thought, but rather that what they are *in fact* talking about is a confused jumble of ideas that includes among its other constituents that determination. So, that the vulgar report upon reflection that this is not the content of their ideas is neither surprising nor informative.

Broughton does go on to cite certain passages in which she finds Hume referring to various specific causal powers in things and referring to secret or unknown powers again in objects.[5] Regarding the specific passages that she cites, I am entirely in agreement. Hume does refer to causal powers, and does understand the science of human nature as discovering these. That, however, is not to say that he takes the science of human nature to discover necessary connections between distinct existences because, of course, Hume does not take causal powers to comprise such necessary connections. More on this in the following two sections.

Like Broughton, Helen Beebee also gives the interpretive thesis that we cannot represent necessary connections among worldly objects relatively short shrift. In contrasting this interpretation of Hume with one according to which Hume holds that we *can* represent such connections, but that we can do so only erroneously, Beebee writes,

> This [the latter] is a better view to attribute to Hume than the other possible view described earlier in this section—the view that our ordinary causal talk does not succeed in *representing* the world at all, so that we are incapable of genuinely *thinking* that one event is necessarily connected to another—because it fits better with what Hume says. For example, he says that the propensity of the mind to spread itself on external object 'is the reason, why we suppose necessity and power to lie in the objects we consider, not in our mind, that considers them' (*T* 167). Hume is here more naturally read as saying that we really *do* literally suppose or think or believe that necessity lies 'in the objects' than read as saying that we cannot, in fact, suppose or think or believe any such thing.[6]

Even if Beebee is right that Hume is in this one instance "more naturally" read as saying that we do literally think that necessity lies in objects, it should be obvious that this hardly constitutes anything more than a modicum of evidence. We have already seen a plethora of other passages where Hume explicitly endorses the conclusion that, "we have no idea of necessity," or that the term "necessity", "has no distinct meaning," so at best the textual evidence would be ambiguous. That said, though, the current interpretation can actually easily accommodate this passage. As we have seen, the vulgar do *take themselves* to represent worldly objects as being necessarily connected to one another, but neither their positing of such worldly objects nor their positing of the necessary connections between these *amounts* to anything. The vulgar use *words*—"objects", "power", "necessity"—in a way that appears to indicate that they are associated with ideas that are different from the ones associated with more mundane terms such as, "perceptions" or "associations", but what Hume discovers is that they are mistaken in this respect. Again, the terms of the vulgar, "make use only of common words, without any clear and determinate ideas." So, when Hume writes that, "we suppose necessity and power to lie in the objects we consider, not in our mind," he is merely speaking with the vulgar, as we have seen him do so many times throughout the *Treatise*. Because the language of the scientist of human nature requires reconceiving what is purportedly represented by the language of the vulgar, in order to communicate his conclusion most effectively, as we have seen, Hume must find a way to state them in *both* languages. So, expressed from within the (inadequate and failed) language of the vulgar, Hume's conclusion is that the necessity that we *think* that we attribute to worldly objects is nothing but a determination of thought. If the term "necessity" refers to anything, it is that.[7] Expressed from within the language of the scientist of human nature, it is that the vulgar represent no such thing as "necessity", but only confusedly bandy about that term, coming closest to picking out that determination of thought, but never really hitting upon it.

Thus, I conclude that the interpretation of Hume's understanding of the relation of the language of the vulgar to that of the scientist of human nature from the previous chapters anticipates and clarifies Hume's conclusion regarding the idea of necessity. It would be nice if something similar worked for Hume's argument for that conclusion, and I believe that it does. Specifically, as we saw in the cases of the ideas of existence, external existence, and body, the account of Hume's use of theoretical representations that I have been defending makes clear the conditions under which such posits are and are not possible. In those previous cases we saw Hume declare such ideas impossible on the grounds that they could not be represented using a simple idea, a complex idea, or perceptible model that is specified to differ from and resemble some explanatory posit in determinate ways. That is, what we discovered in these previous cases was a common argument form. Hume first shows that no simple idea would be adequate to representing the target content, then that no complex idea would, and finally that no theoretical representation would. As noted previously, this final piece of the argument can take either or both of two forms: showing that the target representation can only be specified negatively, as being *entirely unlike* some model idea, or positively, as being *nothing but the same as* some model idea. As we are about to see, Hume's argument concerning the idea of necessary connection take precisely this form.

We can begin with the thesis that we have no simple idea of necessary connection. First, note that Hume holds that, "the terms of *efficacy, agency, power, force, energy, necessity, connexion*, and *productive quality*, are all nearly synonymous" (T 1.3.14.4; SBN 157). Here, then, is a quick argument of just the form that we would expect, for the conclusion that we have no simple idea of power or efficacy, i.e., necessary connection.

> All ideas are, deriv'd from, and represent impressions. We never have any impression, that contains any power or efficacy. We never therefore have any idea of power.
>
> T 1.3.14.11; SBN 161[8]

Here is another case in which Hume's idiom is explicitly representational, and his conclusion clearly indicates his use of both the Copy Principle and the Representational Copy Principle.

(1) "We never have any impression, that contains any power or efficacy."[9]
(2) "All ideas are, deriv'd from [. . .] impressions." (Copy Principle)
(3) We have no idea that is a copy of any impression that contains any power or efficacy. (1, 2)
(4) All ideas [. . .] represent impressions. (Representational Copy Principle)
(5) We never therefore have any idea of power. (3, 4)

As we have noted previously, this straightforward application of the Representational Copy Principle is not alone sufficient to rule out our having an idea of necessary connection (as Hume's conclusion appears to do) because while it demonstrates that we can have no simple idea of power, it leaves open the possibility that we have either a complex idea of it, or can form a theoretical representation of it. We can return to the complex idea in a moment, but before doing so, it is worth noting that just one paragraph after giving this argument against the simple idea of power, following a brief digression concerning whether the mind itself furnishes us with such a simple idea, Hume turns his attention to the suggestion that our simple idea of power, while not experienced directly, can come in the form of a theoretical posit.

> If we be possest, therefore, of any idea of power in general, we must also be able to conceive some particular species of it; and as power cannot subsist alone, but is always regarded as an attribute of some being or existence, we must be able to place this power in some particular being, and conceive that being as endow'd with a real force and energy, by which such a particular effect necessarily results from its operation.
>
> T 1.3.14.13; SBN 161

This approach to the idea of power is exactly in line with the account of substance and theoretical representation that we have already seen Hume give. To represent a property as a substantial property is not to represent it as existing apart from all of the other substantial properties attributed to a given substance, but is rather, "to place this [attribute] in some particular being, and conceive that being as endowed with [it]." To represent a property as a substantial property of a being is to take it and all of the being's other substantial properties to have a common cause in the nature of that being. In the case of power, this would entail taking the substance at hand to be "endow'd with a real force and energy by which such a particular effect necessarily results from its operation." Again, Hume is not here objecting to the notion of substantial explanation *per se*—we have seen him present his own account of this kind of explanation—but only to the specific way in which those who defend the notion of necessary connection deploy it. Hume continues.

> We must distinctly and particularly conceive the connexion betwixt the cause and effect, and be able to pronounce, from a simple view of the one, that it must be follow'd or preceded by the other. This is the true manner of conceiving a particular power in a particular body: and a general idea being impossible without an individual; where the latter is impossible, 'tis certain the former can never exist.
>
> T 1.3.14.13; SBN 161

Again, the target theoretical representation is one in which we represent the connection between the cause and effect as being a necessary one: the posited power would be a power that demonstratively implies the necessity of some other being. As expected, we next find Hume rejecting the possibility of forming any such idea.

> Now nothing is more evident, than that the human mind cannot form such an idea of two objects, as to conceive any connexion betwixt them, or comprehend distinctly that power or efficacy, by which they are united. Such a connexion wou'd amount to a demonstration, and wou'd imply the absolute impossibility for the one object not to follow, or to be conceiv'd not to follow upon the other: Which kind of connexion has already been rejected in all cases.
>
> T 1.3.14.13; SBN 161–2

Hume here refers back to his previous argument, to which we will turn in a moment, that we can form no complex idea of a necessary connection. The gist of that argument is that since no distinct ideas are necessarily connected to one another, given the Representational Copy Principle, no complex idea can represent a necessary connection. Why rehash that argument here? He does so because what Hume considers here is a new twist on that original argument: instead of a standard complex idea, the proposal at hand is to understand power as a theoretical posit. That is signaled in the previous passage by Hume's noting that, "a general idea being impossible without an individual; where the latter is impossible, 'tis certain the former can never exist." As we have seen, the means by which we form theoretical representation is by employing a general term with a uniting principle as its foundation, the use of which term would represent the members of its revival set as bearing some real resemblance to each other, thus marking a genuine ontological difference in the substance of the world.

The problem with using this method to form a theoretical representation of necessary connection is that there is no experience available to serve as a perceptible model for such a posit. Because as Hume notes a necessary connection between distinct existences, "has already been rejected in *all* cases," the only way to form a theoretical representation of necessity would to stipulate that such a connection *differs entirely* from all other relations. As we have seen in a great many cases to this point, however, it is precisely such appeals to a *via negativa* that are Hume's consistent target of criticism. Because Hume has previously argued that it is impossible to form *any* representation of necessity, there is no possible recombination of ideas that can be used to represent a substance as having this feature. The general term "causation" has as its revival set those ideas that are of objects constantly conjoined where one object always precedes the other, and in adopting this term the scientist of human nature adopts the attendant ontological commitment to a real difference between objects so related and objects that do not stand in that relation to one another. There is nothing about that procedure, however, that serves to mark that difference as corresponding to, or reflecting, the difference between objects necessarily connected to each other and those only contingently related. In particular, as we saw in the previous chapter with the idea of continued and distinct existence, there is no way to give determinate content to a representation of necessity at all. That kind of connection "has already been rejected in all cases," so there is no case that can serve as its model.

In fact, Hume puts his conclusion in *precisely* this way a few paragraphs later in considering what the "true idea" of necessity really is.

> Tho' the several resembling instances, which give rise to the idea of power, have no influence on each other, and can never produce any new quality in the object, which can be the *model* of that idea, yet the observation of this resemblance produces a new impression in the mind, which is *its real model*.
>
> T 1.3.14.20; SBN 164–5, emphasis mine

Because we have no impression of the power, influence, or necessary connection between objects, we cannot use such an impression as a model for forming a theoretical representation of necessity more generally. What we can use as a model, however, is the "determination of the mind to pass from one object to its usual attendant," i.e., the inductively-established universal regularity of experience that is the mind's own association of items that are found to be constantly conjoined. While the scientist of human nature represents the difference between causally related items and those that are not so related as the *real* difference between objects constantly conjoined and those that are not, there is no way for the proponent of necessary connection to cast this real difference as determinately reflecting the difference between objects necessarily connected and those that are not. Recall that in the previous chapter we noted that by specifying the referent of a "relative idea" as "whatsoever is the cause of my impressions," Strawson opened himself to the objection that while this *might* refer to continued and distinct existences, for all we know, it could refer to the Deity or nothing at all. Without a way to specify the reference of an idea *determinately*, without a way to give *some* descriptive or representational content to that idea, it is impossible for an idea to refer to anything. Thus, because there is no quality that "can be the model of" the idea of necessity, we cannot represent, or determinately refer, to any such idea.

Recall in this regard the following argument from 1.4.5 in which Hume argues that we cannot conceive of matter as something specifically different from a perception.

> the quality of the object[[. . .] must at least be conceiv'd by the mind; and cou'd not be conceiv'd, unless it were common to an impression; since we have no idea but what is deriv'd from that origin.
>
> T 1.4.5.20; SBN 241–2

In order to attribute any quality to an object, we must at least be able to conceive that quality, and the only qualities that we can conceive are those that are first found in impressions. Thus, if we cannot have any impression that "contains" a necessary connection, we can form no idea of that quality, and so cannot attribute it to any object, even via a theoretical representation. A theoretical representation is a *mixture* of impressions and reason, i.e., it begins with a perceptible model and proceeds by specifying *determinate* ways in which the posited theoretical entity differs from and resembles that model. Since one cannot specify a determinate difference between that which is merely contingently connected and that which is necessarily connected, and one only ever experiences the former, one cannot form any representation of the latter.

Thus, Hume's rejection of the possibility of forming a theoretical representation of necessary connection hinges in an important way on the details of this argument that we can form no complex idea of it. What Hume will need to show is that forming *any* determinate representation of a necessary connection between distinct existences is impossible, and as we saw at the opening of this section, he does take himself to do just that. That argument is the first piece of business to which Hume turns in 1.3.6 "Of the inference from the impression to the idea" after having cleared the decks in the first five sections of Part 3 "Of knowledge and probability" by outlining the place of the idea of causation in our cognitive apparatus, presenting the *differentia* for the relation of causation, and rebutting a few quick attempts to show that the idea of causation comprises the idea of necessity. Finally, Hume turns to his own positive argument that the former cannot include the latter because the idea of necessity is not one that we can form.

> 'Tis easy to observe, that in tracing this relation, the inference we draw from cause to effect, is not deriv'd merely from a survey of these particular objects, and from such a penetration into their essences as may discover the dependance of the one upon the other. There is no object, which implies the existence of any other if we consider these objects in themselves, and never look beyond the ideas which we form of them. Such an inference wou'd amount to [demonstrable] knowledge, and wou'd imply the absolute contradiction and impossibility of conceiving any thing different. But as all distinct ideas are separable, 'tis evident there can be no impossibility of that kind. When we pass from a present impression to the idea of any object, we might possibly have separated the idea from the impression, and have substituted any other idea in its room.
>
> T 1.3.6.1, SBN 86–87[10]

Once again, it will be worth noticing that Hume's argument relies on not only the Copy Principle, but also the Representational Copy Principle. Hume's first premise is that if "the inference we draw from cause to effect [were] deriv'd merely from a survey of these particular objects," it would, "imply the absolute contradiction and impossibility of conceiving any thing different." To "derive" the inference from cause to effect from a survey of objects would be for that inference to *copy* the relations of those objects to each other. If the copied relation is to be a *necessary* connection, then to be a copy of it, the idea representing each of the *relata* would likewise have to be necessarily connected to each other. That is, to represent a necessary connection between distinct objects would require a necessary connection between the distinct ideas representing these. To plug these values into our oft-used schema:

> '*x*' necessarily connected to '*y*' represents *x* necessarily connected to *y*.

To be a copy of a necessary connection, and thereby to represent a necessary connection, the idea of each of the objects represented would themselves have to be necessarily connected. As Hume points out, though, no two distinct ideas are necessarily connected each other. We can always entertain the idea of the cause without also thinking of the effect, and vice versa.

Thus, the representing ideas are not necessarily connected, and so are not copies of any necessary connection, and so do not represent any necessary connection.

Of course, as we have seen, not all complex ideas are copies of their represented complex impressions. The idea of New Jerusalem might perfectly resemble New Jerusalem, but is not caused by any such city. The idea of Paris is caused by Paris, but does not exactly resemble it. So, while Hume shows that no complex idea can be a copy of a necessary connection, it would appear that he has a little more work to do to show that no complex idea can be an idea *of* a necessary connection. This work, however, is easily accomplished. As we noted in our discussion of misrepresentation, when a complex idea misrepresents, what it represents is that the simple impressions that are the objects of its component simple ideas are arranged in the way that those component ideas are arranged in the complex. Since no distinct ideas can be necessarily connected, then no complex idea can represent distinct objects as necessarily connected because to do so would require that its component ideas would be necessarily connected to one another. Thus, Hume's argument here moves from our ability to recombine our distinct ideas, through the impossibility of demonstrative knowledge of the connection between two such ideas, to the impossibility of forming an idea of necessary connection.

So, what we see Hume argue is exactly what we have come to expect. We can have no simple idea of necessary connection. We can have no complex idea of necessary connection. We cannot form a theoretical representation of necessary connection. Having exhausted the possibilities, he finally concludes that we can have no idea of necessary connection at all (and so our idea of causation does not rely on our having an idea of necessary connection). This argument is of roughly the same form that we saw Hume employ in the previous chapter with respect to the idea of external existence, and which we will see him employ in the next chapter with respect to the idea of a substantial soul. In the meantime, however, with the particulars of Hume's argument and conclusion now established, we must return to the puzzles with which we began this chapter: how can Hume's conclusion that we have no idea of necessary connection be compatible with an interpretation according to which Hume understands explanation as proceeding via theoretical posits representing the nature of substance?

Explanation and Expectation

Recall that at the outset of this chapter we noticed an apparent tension between the theses that we can have no idea of necessary connection and that explanation proceeds via theoretical posits representing the nature of substance. The worry here is that substantial explanations cast the observed regularities as *necessarily following from* the nature of the underlying substance. For example, when we explain the fact that gases obey the Ideal Gas

Law by casting them as a collection of molecules themselves obeying familiar mechanical laws, it is plausible to suppose that a gas's being a collection of molecules, etc., *makes it necessary* that it obeys the Ideal Gas Law. That is, it *follows from* a gas being a collection of molecules, and the familiar mechanical laws, that gases *will* obey the Ideal Gas Law. In casting some manifest phenomena as being the result of the nature of some underlying substance, does not substantial explanation imply a necessary connection between the states of both the manifest phenomena and the substance at different times? Or to take the converse problem: if substantial explanation does *not* require necessary connection, and if it is regularities all the way down, so to speak, then is the understanding of substantial explanation attributed to Hume not subject to an vicious regress of explanation, explaining one regularity via an appeal to another, which in turn demands explanation, etc.?

The first thing to note in proceeding is that, as we saw in Chapter 3, for Hume substantial explanation *just is* causal explanation. In accounting for the difference between substances and modes, Hume writes that,

> the difference betwixt these ideas consists in this, that the particular qualities, which form a substance [. . .] are at least suppos'd to be closely and inseparably connected by the relations of contiguity and causation.
>
> T 1.1.6.2; SBN 16

The idea of substance, properly conceived, is the idea of some particular qualities, assigned the same name, that are "suppos'd to be closely and inseparably connected by the relations of contiguity and causation." The relation of contiguity that Hume refers to here is the contiguous appearance of each of the qualities at hand, but that turns out to be a non-starter because as Hume notes a moment later, it is easy enough to imagine a scenario in which a mode displays this same relation.

> The simple ideas of which modes are form'd, either represent qualities, which are not united by contiguity and causation, but are dispersd'd in different subjects; or if they be all united together, the uniting principle is not regarded as the foundation of the complex idea.
>
> T 1.1.6.3; SBN 17

While modes are typically "dispers'd in different subjects," they need not be, and when they are not, what distinguishes a mode from a substance is that with respect to the former, "the uniting principle is not regarded as the *foundation* of the complex idea." "The complex idea" here refers to the general representation that is used to represent a substance, and as we noted earlier, having a uniting principle serve as its foundation is not a feature of general representation *per se*, but rather is a feature unique to the representation of a substance. Specifically, what is implied by the passages above is that the uniting principle that serves as the foundation of a representation of substance is a causal one: to represent a collection of qualities as constituting a substance is to represent those qualities as being causally connected via the nature of some underlying substance. (Hume's example in that section is the malleability, color, solubility in *aqua regia*, etc., of gold.)

The point of recalling all of these details is to notice that, as one would expect, nowhere in Hume's account of substance does *he* mention necessary connection at all. Instead, what he appeals to are the causal relations that are represented in representing substance. Thus, while we should expect there to a be a relation between Hume's understanding of substance and his understanding of causation, Hume himself gives no reason for thinking that his relation will proceed via the representation of necessary connection. What it is to be a substance, for Hume, is *just* to be a collection of qualities united by certain causal relations, and if causal relations do not require necessity, then neither will being a substance. That fact is significant because on one way of motivating the apparent tension between the theses that we can form no idea of necessary connection and that theoretical explanation proceeds via the positing of a substance that underlies the manifest phenomena is itself generated by an implicit assumption that casts causation as comprising necessary connection after all. Here, for example, is one way of formulating an argument to bring out this tension.

(1) Gases are collections of molecules that obey familiar mechanical laws. (Substantial Explanation)
(2) Applying familiar mechanical laws to molecules implies that the state of a gas at one time will be necessarily connected to the state of that gas at another time.
(3) Substantial explanation implies that there are necessary connections between distinct existences. (1, 2)

As should be clear by making this argument explicit, its second step simply takes for granted that familiar mechanical laws represent necessary connections between distinct existences. Thus, in this way of proceeding, the apparent conflict between substantial explanation and the impossibility of an idea of necessary connection is actually generated not by anything particular to substantial explanation, but only by a misconstrual of the nature of familiar mechanical laws. That is, there is nothing about substantial explanation *per se* that requires that such explanation proceed via positing necessary connections. One can take one's preferred understanding of Hume on causation and simply plug that in to substantial/causal explanation and proceed without issue.

What I argue below is that what substantial explanation achieves is to render the inductively-established universal regularity of experience *expected*. We are accustomed to observing macroscopic objects behave according to familiar mechanical laws, and when confronted with a new instance of macroscopic motion, we expect the same. Upon our first noticing that gases obey the Ideal Gas Law, we do not have such an expectation, and so this regularity strikes us as surprising and in need of explanation. In supposing, however, that the behavior of gases is the manifestation of their being composed of microscopic particles that obey the same mechanical laws, we represent this behavior as being of the same kind as that of macroscopic objects, and so our expectation concerning the latter is transmitted to the former.[11] Hume describes this process in his fourth rule by which to judge cause and effect.

> The same cause always produces the same effect, and the same effect never arises but from the same cause. This principle we derive from experience, and is the source of most of our philosophical reasonings. For when by any clear experiment we have discover'd the causes or effects of any phenomenon, we immediately extend our observation to every phenomenon *of the same kind*, without waiting for that constant repetition, from which the first idea of this relation is deriv'd.
>
> T 1.3.15.6; SBN 173, emphasis mine

Probable reasoning, inferences concerning cause and effect, proceed by subsuming ideas under a general term, thereby representing their objects as of a certain kind, and then supposing that all objects of that kind stand in similar causal relations. That is one of the reasons that it is essential to represent theoretical entities via a perceptible model. We have previously observed that this modeling is essential for giving a theoretical representation its *content*, but we can now also note that it is likewise essential to imbuing it with *explanatory force*. It is because theoretical posits resemble that on which they are modeled in determinate ways that they can be represented as of a kind with those objects, and so can be represented as standing in the same causal relations as their model objects. To consider one particularly infamous and striking example, it was reconceiving of celestial objects as of a kind with terrestrial ones that allowed astronomers to apply known laws of terrestrial motion to celestial objects as well.

That the result, if not the end, of causal explanation is forming such an expectation is also implied by the sixth rule.[12]

> The following principle is founded on the same reason. The difference in the effects of two resembling objects must proceed from that particular, in which they differ. For as like causes always produce like effects, when in any instance we find our expectation to be disappointed, we must conclude that this irregularity proceeds from some difference in the causes.
>
> T 1.3.15.8; SBN 174

Causal reasoning generates expectations, and when these expectations are disappointed, causal reasoning also contains a mechanism for correcting itself. As we will see in a moment, it is because causal reasoning has such a self-correcting mechanism in place that that the expectations that at which it ultimately aims are more "fixed," "settled," and "unalterable," unlike those of mere habituation.

That difference is an important one, and to see why consider that it is *repeated* experience of gases obeying the Ideal Gas Law that itself prompts the demand for explanation via the particulate theory of gases. Given that, one might wonder whether that repeated experience alone would be enough to establish the requisite expectation. That is, simply observing the variations in a gas's temperature, pressure, volume, etc., might itself be enough to generate an expectation that, for example, if one raises the temperature of the gas, its pressure or volume will also increase. If that were the case, then either substantial explanation would not be necessary at all or its function could not be to render its *explicandum* expected because that *explicandum* would already itself generate the expectation that is the purpose of such an explanation. That Hume himself considers just such an objection to his account of belief gives us a clue that this understanding of explanation is on the right track; that his answer to that objection fits well with that interpretation is another. First, the objection:

> For it may be said, that if all the parts of that hypothesis be true, *viz. that* these three species of relation are deriv'd from the same principles; *that* their effects in inforcing and inlivening our ideas are the same; and *that* belief is nothing but a more forcible and vivid conception of an idea; it shou'd follow, that that action of the mind may not only be deriv'd from the relation of cause and effect, but also from those of contiguity and resemblance. But as we find by experience, that belief arises only from causation, and that we can draw no inference from one object to another, except they be connected by this relation, we may conclude, that there is some error in that reasoning, which leads us into such difficulties.
> T 1.3.9.2; SBN 107

The objection that Hume considers asks precisely the question that we have posed: why is it that only *causal* explanations generate the expectation that is the aim of explanation, rather than all repeated experience and association? Or another way of putting it: if the end of explanation is rendering an observed phenomena expected, why is it that repeated experience itself and the customs or habits that it generates enough to achieve this end? Here is Hume's answer.

> 'Tis evident, that whatever is present to the memory, striking upon the mind with a vivacity, which resembles an immediate impression, must become of considerable moment in all the operations of the mind, and must easily distinguish itself above the mere fictions of the imagination. Of these impressions or ideas of the memory we form a kind of system, comprehending whatever we remember to have been present, either to our internal perception or senses; and every particular of that system,

> joined to the present impressions, we are pleas'd to call a *reality*. But the mind stops not here. For finding, that with this system of perceptions, there is another connected by custom, or if you will, by the relation of cause or effect, it proceeds to the consideration of their ideas; and as it feels that 'tis in a manner necessarily determin'd to view these particular ideas, and that the custom or relation, by which it is determin'd, admits not of the least change, it forms them into a new system, which it likewise dignifies with the title of *realities*.
>
> T 1.3.9.3; SBN 107–8

The most direct way to form a belief is via the employment of memory or sensation. Impressions naturally have the highest degree of force and vivacity, and since memories are intended to be exact copies of these, they retain a high degree of this force and vivacity. Matters are less direct when it comes to causal reasoning because as we have seen reason is a recombinatory faculty, and so will not retain as high a degree of force and vivacity as memory does. Still, because the mind finds that causal reasoning "admits not of the least change," the products of that faculty retain enough force and vivacity to produce belief.

The question now is what exactly Hume means in claiming that causal reasoning, "admits not of the least change," and my suggestion will be that—as per my earlier discussion of reason—what distinguishes the influence that reason has on belief is that its products are *reliable* in a way that those of the imagination are not. That is, what Hume means by, "admits not of the least change," is that reason's representations, when successful, produce expectations that are satisfied by further experience and therefore do not stand in need of revision.[13] That contrasts with the representations of the imagination (or custom, habit, and education) insofar as those faculties produce expectations that are all too often thwarted or disappointment by experience. It is because reason functions to explain the manifest phenomena that its representations are in accordance with experience, and so prompt a kind of meta-expectation-formation: we come to expect that the expectations generated by reason will be confirmed, whereas those generated by the imagination will be disconfirmed, and so the object-level expectations of reason come to be stronger than those of the imagination.[14]

Hume continues presenting his contrast between causal reasoning and the imagination by considering the example of his belief in his idea of ancient Rome, and gives the following account of that belief.

> All this, and every thing else, which I believe, are nothing but ideas; tho' by their force and settled order, arising from custom and the relation of cause and effect, they distinguish themselves from the other ideas, which are merely the offspring of the imagination.
>
> T 1.3.9.4; SBN 108

Hume notes that the force of the ideas that are the product of causal reasoning is one way in which they distinguish themselves from the ideas of the imagination, but of course it is this very force that we are aiming to understand. Thus, the other distinguishing feature that Hume mentions here is particularly important: the "settled order" of these ideas. This order cannot be merely the order internal to Hume's complex idea of Rome because, for example, his idea of New Jerusalem, which is a product of the imagination and in which Hume does not believe, has a similar internal order.[15] Instead, what distinguishes Hume's idea of Rome is the fact that this internal order is specifically a *settled* one. When he thinks of Rome, and matches this idea to the "conversation and books of travelers and historians" (T 1.3.9.4, SBN 108), he finds his idea is restrained in a way similar to the way that memory is restrained: there is an (indirect) experiential check on the form that his idea can take, and so his idea (unlike those of the imagination) is settled insofar as it must stay within the bounds of this constraint. In order to accord with these accounts, it must remain constant and "admit not of the least change."

Continuing to discuss the contrast between causal relations and the other associations of ideas, Hume notes that,

> There is no manner of necessity for the mind to feign any resembling and contiguous objects; and if it feigns such, there is as little necessity for it always to confine itself to the same, without any difference or variation. And indeed such a fiction is founded on so little reason, that nothing but pure *caprice* can determine the mind to form it; and that principle being fluctuating and uncertain, 'tis impossible it can ever operate with any considerable degree of force and constancy. The mind forsees and anticipates the change; and even from the very first instant feels the looseness of its actions, and the weak hold it has of its objects.
> T 1.3.9.6; SBN 109–110

While we might *feign* that the deliverances of the imagination are real, while we might entertain such a *fiction*, such pretending is inherently unstable. It is only "pure caprice" that would prompt us to such action in the first place, and such whims cannot "ever operate with any considerable degree of force and vivacity." Here again is that meta-expectation: while such a fiction would generate an expectation that resembling and contiguous objects should continue to be experienced as such, the mind has been trained to be distrustful of the "fluctuating and uncertain" caprice of the imagination, and so abandons that faculty as an unreliable mechanism. By contrast,

> The relation of cause and effect has all the opposite advantages. The objects it presents are fixt and unalterable. The impressions of the memory never change in any considerable degree; and each impression draws along with it a precise idea, which takes its place in the

> imagination, as something solid and real, certain and invariable. The thought is always determin'd to pass from the impression to the idea, and from that particular impression to that particular idea, without any choice or hesitation.
>
> T 1.3.9.7; SBN 110

The objects presented by causal reasoning are "fixt and unalterable," and are drawn along by the impressions of the memory. That is, causal reasoning reliably tracks our experience because it is aimed at explaining, and is thereby ultimately accountable to, that experience. It is because of this answerability to experience that the expectations formed by reason's theoretical posits prove to be reliable and therefore to generate stable meta-expectations. When successful the theoretical posits of reason represent something "solid and real, certain and invariable," and so track and explain their target manifest phenomena. As such, "thought is always determin'd to pass from the impression to the idea [. . .] without any choice or hesitation." That is, when we experience some manifest phenomenon for which reason has successfully posited as an explanation some theoretical entity underlying it, our minds move naturally from this manifest phenomena to the idea of that which is posited as explaining it. Such customs are reinforced by further confirming evidence, and so reason proves itself reliable in a way that imagination cannot. Thus, the deliverances of reason can be and are trusted in a way that those of imagination are not.

Back to Gases

We repeatedly observe the behavior of gases and arrive at an inductively-established universal generalization concerning that behavior: the Ideal Gas Law. This regularity of experience is not itself enough to generate a robust expectation of gases continuing to behave in this way because thus far it is merely an observed relation of contiguous and resembling items. The gas in this balloon was spatially contiguous with this heat source, which relation itself was temporally contiguous with a gas in a balloon of a greater volume, etc. Such experiences might generate a weak expectation of similar conjunctions, but nothing more than that. When, however, we offer a *causal explanation* of these events, when we subsume gases under the general term "particles" (or "particles obeying familiar mechanical laws"), then a steadier and more reliable expectation is formed. In representing a gas as a collection of molecules obeying familiar mechanical laws, we form the expectation that gases will exhibit the same regularities that macroscopic objects do because we reconceive of gases as being of the same kind as such objects, and so the expectation from the familiar case is applied to the posited one. The end of causal/substantial explanation is the firm and stable expectation, produced only by causal reasoning, that inductively-established empirical regularities will continue. It is by appealing to the nature, powers, and essence of the substance underlying manifest phenomena that causal explanations explain why the particulars obey the empirical laws that they do and thereby generate the fixed and unalterable expectation that they will continue to do so. So, the argument from earlier should proceed as follows.

(1) Gases are collections of molecules that obey familiar mechanical laws. (Substantial explanation)
(2) Applying familiar mechanical laws to molecules renders the state of a gas at one time expected given the state of that gas at another time.
(3) Substantial explanation renders expected some phenomena given some other one. (1, 2)

As should be obvious, this conclusion is not one that should bother a version of Hume who takes the end of explanation to be rendering a given inductively-established universal regularity expected. Thus, the apparent tension that we observed at the outset of this chapter is resolved. Positing a substance to explain inductively-established universal regularities does not conflict with the thesis that we can have no idea of necessary connection because the idea of substance, while essentially causal, like other causal relations for Hume, does not include the idea of necessity. Instead, a substantial/causal explanation renders its *explicandum* expected by positing an object that is of the same kind as some antecedently familiar one and thus transferring the fixed and unalterable expectation regarding the behavior of the latter to the former. Nothing about this procedure requires that we represent the causal relations governing underlying substances as any more necessary than those governing manifest phenomena, and nonetheless it does account for why the result of such a procedure is a more stable expectation than is formed by repeated observation alone.

With that said, it is worth considering a concern with the understanding of Hume just proposed. According the interpretation of Hume that I have been presenting, the *explicandum* of a substantial explanation is not the particulars themselves, but rather the fact that the particulars obey the inductively-established universal regularities that they do. This point is significant because if substantial explanations do not involve positing a *necessary* connection between the states of certain substantial properties, then it might appear that the only alternative is to revert to an account of explanation as consisting in nothing more than the subsumption of observed phenomena under such increasingly general inductively-established universal regularities themselves. That is, if what a substantial explanation does is render its *explicandum* expected by representing it as of a kind with some more familiar phenomenon that is itself expected to behave a certain way, then it appears that substantial explanation does nothing more than subsume some phenomena under the very same inductively-established universal regularities that govern some other phenomena. For example, in casting gases as consisting of molecules that obey familiar mechanical laws, this explanation seems to merely derive the Ideal Gas Law as an instance of a more general regularity (the familiar mechanical laws). If this is all substantial explanation accomplishes, then Hume appears to have a less robust understanding of scientific explanation than I have been supposing.

Recall that in Chapter 2 I contrasted my interpretation of Hume as a scientific realist with that of De Pierris who casts Hume as an inductivist. As we saw then, the central pillar on which De Pierris builds her interpretation of Hume is her understanding of Hume's use of "inductive proof."

> Hume's notion of inductive proof, which is at the heart of his conception of causation and scientific methodology, consists in a universalization [. . .] of our past and present uniform experience, with the attendant assumption that nature is, in Newton's words, "ever consonant with itself"[16]

On De Pierris's line, scientific explanation consists entirely of such inductive proofs, of discovering increasingly general inductively-established descriptions that also most accurately picture manifest phenomena. By contrast, on the line that I have been defending, while Hume certainly holds that whatever explanation we formulate must be consonant with manifest phenomena, I have urged that this is compatible with reading Hume as a scientific realist who understands explanation as aiming at discovering the nature, powers, and essence of the substance underlying those manifest phenomena. For that to be the case, it must be that Hume understands scientific explanation at aiming at more than merely discovering maximally universal generalizations because, as De Pierris points out, such generalizations do not represent the hidden microstructure of the world at all, but concern *only* manifest phenomena. If, however, the proximate end of explanation is to form a stable *expectation* (which itself, of course, concerns only manifest phenomena), and such expectations are formed by representing some manifest phenomena (e.g., a gas obeying the Ideal Gas Law) as of a kind with some other phenomena that has already been "inductively explained" in order to subsume the former under the same generalizations as the latter, then it looks like De Pierris's picture is exactly right. Explanation consists of subsuming more and more manifest phenomena under a maximally universal set of inductively-established generalizations.

The game is not lost quite so easily, though, because this way of casting the matter overlooks an essential element of the way that the example at hand functions. The *way* that a gas obeying the Ideal Gas Law is subsumed under the same inductively-established universal regularities (familiar mechanical laws) is not itself through induction alone. That is, while the Ideal Gas Law is arrived at inductively, the *explanation* for why the Ideal Gas Law holds is not itself established via induction, but rather by *reconceiving* gases as collections of molecules. That reconception, though, requires moving *beyond* the manifest phenomena by deploying an ontologically-committing theoretical posit. As we saw above, in considering Hume's fourth rule by which to judge cause and effect, the behavior of gases is only subsumed under familiar mechanical laws by understanding gases as being of the same kind as macroscopic objects, and doing this requires representing gases as a collection of unobservable molecules.

This fact is no quirk of the chosen example either. Consider one of the other cases that we have had occasion to examine: the explanation of the substantial properties of gold via an appeal to gold's atomic structure. Here the idea would be something like the following. The *explicandum* at hand in that example is gold's having certain substantial properties and exhibiting the attendant empirical regularities, e.g., dissolving in *aqua regia*. As we saw Hume note, the explanation for the conglomerate substantial properties of gold will make an appeal to the causal powers of the substance underlying these manifest properties, and in considering this example we saw that representing gold as a collection of atoms with an atomic number of 79, etc., fits this bill nicely. What we can now add to that picture is that it is by taking gold to consist of atoms with such properties that we come to render the observed regularities in the behavior of gold expected. This stable expectation, though, is not the mere subsumption of gold under some previously familiar generality, but crucially depends on conceiving the observable properties of gold as a manifestation of its unobservable atomic properties.

This point illustrates the difference between what I have been calling substantial explanation and Hume's conception of argument from analogy more generally. Arguments from analogy for Hume are arguments that infer like effects from like causes or like causes from like effects.[17] While such arguments have their places, it is important to note the difference between these and substantial explanation. While a substantial explanation certainly requires taking like causes to produce like effects and vice versa, the explanatory force of a substantial explanation derives precisely from its ability to cast these constant conjunctions in terms of the nature and powers of the substance underlying them. Thus, we find Hume, by way of Philo, in Part IV of the *Dialogues*, arguing that casting God as a divine intelligence does nothing to *explain* the order and complexity of the universe because in doing so we do nothing more than take God's mind to have a similar, albeit mental, complexity and order that then demands its own explanation in turn (D 4.8–12). Philo's thought here is that merely positing like causes for like effects does not make explanatory progress because insofar as the observed effects demand explanation so too will the posited causes. As he notes, a better approach to explaining the observed regularities in the universe is, "to say, that the parts of the material world fall into order, *of themselves, and by their own nature*" (D 4.10, emphasis mine). Since an explanatory appeal to "more of the same" is bound to result in an infinite regress of explanation, and so fail—because it will "excite an inquisitive humour, which it is impossible to ever satisfy" (D 4.9)—the proper place to seek an explanation is not in a *mere* argument from analogy, but rather in the *nature of the phenomena at hand* themselves. While the representation of that nature will certainly involve a kind of analogy—insofar as theoretical representations are specified as differing and resembling their perceptible models in determinate ways—it will do more than merely attribute like effects to like causes and vice versa. It will additionally require reconceiving the observed phenomena as the manifestation of the nature and powers of the substance underlying it.[18]

We also see both of these senses of "analogy" at work in Hume's discussion of the Deity in the *Enquiry*.

> The Deity is known to us only by his productions, and is a single being in the universe, not comprehended under any species or genus, from whose experienced attributes or qualities, we can, by analogy, infer any attribute or quality in him.
>
> EHU 11.26; SBN 144–145

Insofar as we can conceive of the Deity at all it is merely as the cause of the universe, which requires the kind of argument from analogy described above. "As the universe shews wisdom and goodness, we infer wisdom and goodness. As it shews a particular degree of these perfections, we infer a particular degree of them, precisely adapted to the effect which we examine" (EHU 11.26; SBN 144–5). Applying the argument from the *Dialogues* to this inference, we should conclude that supposing the Deity to have such degrees of wisdom and goodness will not *explain* the wisdom and goodness of found in the world. To gain explanatory purchase, we would have to identify something about the *nature* of the Deity (an attribute or quality) that would account for this wisdom and goodness and could halt the explanatory regress that they initially seem to prompt. Because, however, the Deity is not of the same *kind* as anything that we find in experience, because our representation of the Deity has no perceptible model, we cannot use an analogical *representation* to attribute anything to its nature. Arguments from analogy, which merely attribute like effects to like causes and vice versa, are distinct from *representations* from analogy, which require specifying determinate resemblances and differences between a posited theoretical entity and some perceptible model. It is only the latter that locate the source of the regularities of experience in the nature of a substance itself, and it is that that is required to end the regress of explanation that inductively-established universal regularities of experience first prompt.

Of course, this supposes that attributing a certain feature to the nature of a posited substance affords some relief from explanatory demands that merely observing that feature in some manifest phenomena does not, and at first blush that might seem like a strange suggestion. For example, one reason for thinking that attributing a regularity to the "nature" of an object provides explanatory relief is because in doing so one takes that regularity to be a *necessary* feature of such substances. As should be clear, that is not a line that is open to Hume as I have been interpreting him. So, what does attributing a regularity to the nature of a substance accomplish? To answer that question, consider the following passage from Hume's investigation into probability.

> *This* dye form'd as above, contains three circumstances worthy of our attention. *First*, Certain causes, such as gravity, solidity, a cubical figure, *&c.* which determine it to fall, to preserve its form in its fall, and to turn up one of its sides. *Secondly*, A certain number of sides, which are suppos'd indifferent. *Thirdly*, A certain figure, inscrib'd on each side. These three particulars form the whole nature of the dye, so far as relates to our present purpose; and consequently are the only circumstances regarded by the mind in its forming a judgment concerning the result of such a throw.
>
> T 1.3.11.10, SBN 128

What is interesting about this passage is that Hume takes the three facts about the die that he lists to "form the whole nature of the dye, *so far as relates to our present purpose*." Clearly, what he has in mind is that these are all the facts that are relevant to the consideration of probability that he is undertaking. That is interesting because it implies that there is at least one way of talking about the nature of a thing in which that nature is not an *intrinsic* feature of the object, but rather is constituted relative to the purposes of the person considering it. It might be that the die has a nature of its own, but *insofar as it is relevant to a particular inquiry* its nature consists of those properties that are relevant to that inquiry. What *matters* about a die in this case is that certain forces cause to fall on only one side, all of the sides are equally likely to land face up, and that each side has a number on it. Thus, it is these facts, in the context of Hume's particular interest in probability, that constitute the die's nature. These are all the facts that are required to *explain* the result of the throw of the die. In another context, e.g., one in which one was investigating the relative buoyancies of ivory and plastic, a different feature of the die would be relevant, and so one could stipulate a different nature for it in that context. Most importantly, though, in the context of the investigation of probability a demand for an explanation of the fact that a die will always land face up on one side or another is inappropriate. First, that explanatory demand is outside the scope of the investigation being undertaken, but also for the purposes of that investigation that fact is one that is taken to *belong to the nature* of the die: *ex hypothesi* a die will land face up on one side or another.

To see how this way of thinking about the nature of an object is germane to the present threat of explanatory regress, we can return to the example of explaining a gas obeying the Ideal Gas Law via casting it as a collection of molecules obeying familiar mechanical laws. What we have been wondering is why it is that casting the regularity that is a gas obeying that law as being the result of its being composed of molecules themselves subject to a different, but still merely contingent regularity, would be explanatory. My suggestion has been that this explanatory force stems from locating the regularity of the familiar mechanical laws in *the nature* of molecules. What I think the case of the die brings out is that while doing this does not preclude asking after an explanation for the fact that molecules obey familiar mechanical laws *per se*, it does preclude asking that question *in the context of explaining a gas obeying the Ideal Gas Law*. For the purposes of that latter explanatory endeavor, we attribute these motions to molecules *as part of their nature*, thereby postponing any questions about why they behave in that way. In this way, we achieve at least a provisional explanation. When we ask why the die landed on six, we appeal to the nature of the die as Hume explains it for our answer. When we ask why a gas obeys the Ideal Gas Law, we appeal to the nature of the gas qua a collection of molecules obeying familiar mechanical laws for our answer. That we do not also explain why a die is as it is or why the familiar mechanical laws are as they are is not germane to the success of those explanations. Such further explanatory demands are forestalled precisely by locating these explanations as stemming from the nature of the substance at hand. Notice that an argument from analogy that posits "more of the same" fails to accomplish even such a provisional explanation. Appealing to a complex and intricate mind to explain the complexity and intricacy of the universe does nothing. Appealing to the die having landed on six in the past does nothing. Appealing to further facts about the *nature* of these substances—that the matter of the universe contains a principle of motion within itself, or that the die is evenly weighted, etc.—does move the explanatory progress along by at least one degree, and points to the proper place for inquiry to continue.

Here is one way to think about this. In order to form a theoretical representation, one must specify determinate ways that the posited theoretical entity both resembles and differs from some perceptible model. To make things simple, suppose we represent molecules by stipulating that they are like billiard balls insofar as they obey familiar mechanical laws, but unlike them insofar as they are much smaller. In that case, obeying familiar mechanical laws is *part of the content of the representation* "molecules". As in the case of the die, it is true *ex hypothesi* that molecules obey familiar mechanical laws; their doing so forms, "the whole nature of the [molecule], so far as relates to our experience." Most importantly, in the context of the investigation of the behavior of gases, a demand for an explanation of the fact that molecules obey familiar mechanical laws is inappropriate.

To bring this back to the broader interpretive issues at hand, consider the link that Hume draws between what is "natural and essential to any thing" and what is expected of that thing in the following passage.

> We may remark, that tho' in a succession of related objects, it be in a manner requisite, that the change of parts be not sudden nor entire, in

> order to preserve the identity, yet where the objects are in their nature changeable and inconstant, we admit of a more sudden transition, than wou'd otherwise be consistent with that relation. Thus as *the nature of a river consists in the motion and change of parts*; tho' in less than four and twenty hours these be totally alter'd; this hinders not the river from continuing the same during several ages. *What is natural and essential to any thing is, in a manner, expected; and what is expected makes less impression, and appears of less moment, than what is unusual and extraordinary.* A considerable change of the former kind seems really less to the imagination, than the most trivial alteration of the latter; and by breaking less the continuity of the thought, has less influence in destroying the identity.
>
> T 1.4.6.14; SBN 258, emphasis mine

When we take a feature of an object to be part of its nature or essence, we are not surprised to observe that object having that feature, whereas another object's exhibiting the same feature will be surprising if that feature is not part of its nature or essence. It is a happy coincidence that Hume specifically cites the motion and change of parts of a river as his example. Because we take this motion to be due to the nature of being a river, it is entirely expected and needs no explanation. Should we observe an exactly similar motion in an object that does not contain that motion as part of its nature or essence, such a demand for explanation would arise. For example, if one were to stumble upon a video of a bridge wobbling up and own and side to side, since this behavior would be surprising and unexpected, a demand for an explanation—earthquake!—would be entirely appropriate. So, again we see that casting a certain behavior as part of the nature or essence of an object renders that behavior expected *of that thing*, even if that same behavior demands explanation when exhibited by a different object.

That macroscopic objects obey familiar mechanical laws is surprising and thus in need of explanation. That molecules obey familiar mechanical laws is *stipulated* as part of the explanation of the behavior of gases and so needs no explanation (at least at first). Thus, that we cast a gas as a collection of molecules whose nature or essence is to obey familiar mechanical laws forestalls the demand for explanation of these regularities, at least in this context, in a way that it does not in the context of observing macroscopic objects to obey those same laws. It is one thing to simply *observe* a regularity—that can be surprising, unexpected, and demand explanation—whereas it is another thing entirely to *postulate* such a regularity as being in the nature of a theoretical entity—in that case the regularity is neither surprising, unexpected, or in need of explanation, but rather it is part of the very explanatory hypothesis itself (although what is so stipulated can come to require explanation in a difference context).

We can now return to the regress of explanation that was the second way of framing the tension to be addressed in this section. If it is the inductively-established universal regularities of experience that demand explanation, and this explanation cannot make appeal to the necessary connections between objects in accounting for such regularities, then it would seem that what will do this explaining will be the regularities governing the underlying substance. So, it is regularities all the way down. If that is right, though, then the regularities of substance appealed to as the *explanans* of the regularities of experience will themselves stand in need of explanation, etc. Thus, we set forth on a path that can only lead to an infinite regress, a circular explanation, or an arbitrary stopping point.

What we can now see is that this way of framing the global situation concerning explanation ignores the importance that context plays in local explanations. In the context of explaining the behavior of gases, reconceiving them as collections of molecules obeying familiar mechanical laws, and thereby attributing these motions to the very nature of molecules, forestalls a demand to explain that motion insofar as it is part of the very content of the representation of this theoretical substance that it have this feature. Whereas in other contexts it will be perfectly appropriate to demand an explanation for this regularity in its turn, in the context of explaining the phenomenon at hand such a demand is out of place. In positing a certain substance as underlying the regularities at hand, one must model this substance on some manifest phenomena, and in so doing, part of the very content of the theoretical representation formed will be that the posited substance resembles that model in some way, shares some regularities with it. Thus, while the regularities of the manifest phenomenon might demand explanation (because they are unexpected or surprising), the same regularities of the posited phenomena are not only expected and unsurprising in such a context, but are so precisely because they are, in that context, stipulated to be the very nature of the posited substance.

So, what differentiates my interpretation of Hume's understanding of the science of human nature from that of De Pierris, and puts at least a temporary halt to the regress of explanation, is the explanatory role that I accord to theoretical posits. The aim of explanation is to render the inductively-established universal regularities of experience *expected*, and thereby to most accurately mirror the real structure of the substance of the world. While such regularities themselves produce a kind of weak expectation via the associations that their repeated experience causes us to form, such expectations are not *stable*, whereas those that proceed via substantial explanation are. Such posits attempt to mark a real difference in the ontological makeup of the world and thus aim at a more robust tracking of experience. In forming representations of them, we attribute to them a nature that both explains the phenomena that are their *explicandum* and which at least temporarily satiates our desire for explanation.

With that said, we have now assembled the last pieces needed to return to the framing puzzle presented in Chapter 1 to offer a solution it, and that will be the business of the next section.

Representing the Substance of the Mind: A Puzzle Solved

With this clarification regarding the nature of substantial explanation in hand we are now finally in a position to offer a solution to the puzzle raised in Chapter 1, one that will draw on all of the work completed thus far. We can begin, as in Chapter 1, with the case of the simple-complex distinction. I there began with the observation that since our phenomenology is always and everywhere complex, the revival set for the general term "simple idea" consists of only complex ideas. That fact becomes problematic when combined with the thesis that a general term represents nothing more nor less than the objects of the members of its revival set as resembling each other. What these two theses imply is that the content of the general term "simple idea" is that certain complex ideas resemble each other. That conclusion, in turn, would spell trouble for Hume because the explanatory force of the simple-complex distinction derives from its ability to be leveraged in explaining such inductively-established universal regularities as the human ability to form a great variety of novel thoughts that are systematically connected to a relative paucity of actual experience. That our complex ideas are composed of *simple* ideas explains that fact; that our complex ideas are composed of the subset of complex ideas that constitute the revival set of "simple idea" does not. So, the puzzle is how to account for Hume's commitment to taking the representative content of general terms such as "simple idea" as representing something more than merely the resemblance of the objects represented by the members of their revival sets.

Having come a very long way from the formation of that puzzle, and gathered a plentiful variety of interpretive components as we have proceeded, it will be helpful to list these briefly before bringing them together to make our solution. So, here are the theses that will together solve this puzzle.

a) Reason discovers via induction the universal regularities of experience that demand explanation.
(Ch. 2)

b) Reason also contributes to forming the theoretical representations that explain these insofar as it is a recombinatory faculty the function of which is to explain.
(Ch. 2 & 3)

c) As an explanatory faculty, reason, is answerable to experience, and is therefore (comparatively) reliable, which in turn makes the expectations it engenders more stable than those produced by observation alone.
(Ch. 3)

d) The content of a theoretical representation is constituted by the determinate ways that it is specified as resembling and differing from some antecedently contentful representation (its perceptible model).
(Ch 2)

e) It is because theoretical posits resemble that on which they are modeled in determinate ways that they can be represented as of a kind with those objects, and so can be represented as standing in the same causal relations as their model objects.

(Ch. 5)

f) The science of human nature aims to correctly represent the substance of the human mind.

(Ch. 3)

g) We represent substance by employing a general term with a uniting principle as its foundation.

(Ch. 3)

h) The uniting principle so employed represents the manifest properties that are the substantial properties of an object as stemming from a common cause.

(Ch. 3)

i) Theoretical posits explain the inductively-established universal regularities of experience by casting these phenomena as the manifestation of the causal powers of the substance underlying them, thus including these phenomena in the revival set of the general term that includes this underlying substance, thereby rendering them expected.

(Ch. 5)

j) It is by replacing a predecessor language with a successor one that has greater explanatory force that we represent the resemblance relations pictured by the latter as more real than those pictured by the former.

That is a lot of moving parts, but what I hope to show is that they move together in harmony to provide a solution to our framing puzzle. To that end, by way of an example, I will use these theses to account for the representational content and explanatory force of, and ontological commitment to, the simple-complex distinction.

That account begins with reason's discovering via induction a certain universal regularity in experience, in this case the systematic connectedness of our ideas with our impressions (a). The explanation of these regularities takes the form of an appeal to the nature of the substance underlying them, in this case to the simple components that constitute our complex phenomenology (f). To represent this substance, the scientist of human nature begins with some perceptible model (complex ideas) and specifies the determinate ways that the theoretical posit at hand resembles and differs from that model: simple ideas are, like complex ideas, images, but unlike complex ideas cannot be "distinguished" into parts (d). Thus, the complex phenomenology found in experience is recombined by reason to form a representation with novel theoretical content: e.g., where we previously associated the taste, smell, and color of an apple solely with each other via relations of contiguity, these each now come to be associated with each other via a resemblance relation that also includes associations with other "simple ideas" such as a dot on a piece of paper, etc. (b).[19]

Note that this novel content is still only that certain complex ideas resemble each other, and so not yet the full-fledged representation of genuinely simple ideas at which we are aiming. To earn this latter, we must employ a mechanism for moving beyond the content of a mere general term: we must represent that content as itself reflecting some real ontological distinction in the substance of the human mind. Recall, then, that general terms can be used to represent either modes or substances, and that it is only when a general term has a "uniting principle as its foundation" that it represents the latter (g). Thus, in order for the term "simple idea" to represent more than merely the resemblance of the objects of the ideas in its revival set, it will have to have such a uniting principle. This uniting principle is that the term "simple idea" must represent the manifest properties that it is meant to explain as all stemming from a common cause (h), which we accomplish by adopting the term "simple idea" (and abandoning competitor general terms), thus casting the objects represented by the members of the revival set "simple idea" as *really* resembling each other, and thereby as constituting a distinct ontological kind with its own nature and causal powers (j).

Of course, it is important to note that the representation hereby formed is not what one might have expected or hoped for at the outset. It is one of the limits that Hume deliberately puts on any theory of representation that its pictorial character (and also therefore its representational content) must consist of an arrangement of simple ideas themselves copied from some simple impression. Thus, if what one expects to find in a theoretical representation is, for example, a representation of *just* a simple idea, or a perfect picture of the internal structure of an atom, or a mental movie of the movements of the molecules that account for a gas obeying the Ideal Gas Law, one will of necessity be disappointed. The only means by which we can represent such theoretical entities is via the recombination of the images that constitute the representational media of the mind. We simply cannot *picture* an atom, or a collection of molecules, or a simple idea in our mind's eye.[20] What we can do, though, is take certain of the pictures that we do form—e.g., of the members of the revival set of "simple idea" as resembling each other—as reflective of more than just our own associative tendencies. In adopting certain but not other general terms into our language, we can take the resemblance of the objects of their members as being *real* in a way that attempts to carve nature at its joints, so to speak. To put it into a slogan: even if we cannot *picture* the substance of the mind, at least we can *mirror* it. That is, we can arrange our representations in such a way as to reflect the real structure of the mind, even if we cannot get as far as forming an image of its substance. By combining Hume's familiar imagistic theory of mental representation with a mechanism for expressing ontological commitment (via the replacement of one set of images by another), we can earn for Hume the explanatory force of successful scientific investigation to which he repeatedly commits himself.

Before turning to accounting for the explanatory force of these ontological commitments via the interpretive theses remaining from the above list, it will be worth pausing here to resolve one more apparent tension in the current interpretation. Throughout this study, I have repeatedly relied on the schema

> 'x'R'y' represents xRy

to capture Hume's theory of mental representation. One aspect of the account that I have just been summarizing is that we represent a substance as the common cause of its substantial properties, and that we do so by adopting the general term for that substance, thereby representing all instances of that substance as really resembling one another in a way that attempts to carve nature at its joints. That proposal, however, seems to conflict with a commitment to universal scope of the above schema insofar as one would expect that representing a substance as the common cause of its substantial properties would be accomplished via a simple instantiation of that schema, not this more elaborate mechanism of replacing one set of general terms with another. Why are substances not represented, for example, like so:

> "substance" causes "substantial property" represents substance causes substantial property?

To answer that question recall that we noted in our discussion of 1.1.6 that Hume rejects the notion of a substance as that which in no way resembles any observable property. At the time, I noted that this does not itself imply that substance is identical to some observable property or properties, but also leaves open the possibility of a notion of substance as that which differs in some ways, but also resembles some such property or properties. What we can see now is that even when properly conceived, the idea of a substance still does not represent its object *directly*. What specifying the determinate ways that a theoretical posit resembles and differs from its object does is allow for the formation of a determinate revival set for a general term the purpose of which is to mark a real difference in the ontological makeup of the mind. What it does not do is *picture* its own special object. This is the way that theoretical representations differ from straightforward complex representations: the latter are *images* of their objects, whereas the former are not. A theoretical representation consists of images, and imagistically represents its objects as resembling each other, but unlike straightforward complex ideas, this does not exhaust its representational content. In adopting the general terms of the science of human nature, we take the resemblance between the objects of the members of its revival set to be really different in a way that outstrips the images that we form of them. Again, we take these distinctions to *reflect* a real difference in the substance underlying the manifest phenomena, even if we cannot *picture* this difference.

Thus, in the schema above, there is no mental image that could appear in the place of "substance", and so this schema proves inadequate to representing the causal structure of the substance underlying experience. One might be tempted to substitute in the general term for the substance at hand, but as I have argued, by itself that general term represents only that the objects of its members resemble each other, and that fact is not sufficient to generate the explanatory power necessary to fuel the explanations of the science of human nature. That is why the mechanism of representing real differences in substances via the replacement of one set of general terms with another proves necessary (and successful). We represent these causal powers by representing collections of substantial properties as resembling one another (e.g., by including contiguous occurrences of yellow, fusibility, malleability, etc., in the reveal set of "gold") and then representing this resemblance as marking a real distinction in the substance of the mind.

That said, forming such a representation nonetheless does represent the specific causal powers of the substance indirectly represented, and doing so is necessary for that theoretical posit to playing the explanatory role that it does. It is by subsuming certain substantial properties under a novel combination of general terms that the substance so represented is represented as having distinct causal powers (e). To see how this works for "simple idea" we can once again build the revival set of that term from the ground up. As Hume tells us at 1.1.1.2, to form the revival set of "simple idea" one must distinguish one's complex ideas into their component parts. This can be done is a number of ways. The first example Hume gives is distinguishing the taste of an apple from its color. To understand what Hume has in mind there, we can look instead at the procedure for distinguishing the color of a globe from its shape, which is similar in kind, but more explicit on exactly what distinguishing is. Specifically, as we saw in Chapter 1, in the latter case Hume makes clear that one does not distinguish the color and shape of a globe by perceiving each separately, but rather by including the globe among the revival sets of the two general terms "white" and "globular".

> Thus when a globe of white marble is presented, we receive only the impression of a white colour dispos'd in a certain form, nor are we able to separate and distinguish the colour from the form. But observing afterwards a globe of black marble and a cube of white, and comparing them with our former object, *we find two separate resemblances*, in what formerly seem'd, and really is, perfectly inseparable. After a little more practice of this kind, we begin to distinguish the figure from the colour by a distinction of reason; that is, we consider the figure and the colour together, since they are in effect the same and undistinguishable;

> but still view them in different aspects, *according to the resemblances, of which they are susceptible*.
>
> T 1.1.7.9; SBN 21–2, emphasis added

As I highlighted when last we encountered this passage, it occurs in the context of Hume's using his theory of general representation to account for the distinction of reasons such as that between a globe's shape and its color. As I have emphasized, Hume's theory of general representation itself relies on the resemblances that we find between objects, and that is precisely what he leverages here. We distinguish the shape and color of the globe by using the theory of general representation to represent the globe as, on the one hand, resembling other globular objects, and other white object on the other.

Likewise in the case of the apple, then, we distinguish the taste and color of the apple by including the apple in the revival sets of the two general terms "apple-taste" and "green". Similarly, in the case of the disappearing dot, we can distinguish the dot from the piece of paper on which it appears by associating that tableau at once with certain resembling images such as those of a blank piece of paper on the one hand, and those of a single dot on a clay tablet on the other. Once a sufficient number of these revival sets are formed, we can form the revival set of "simple idea" by associating each of these revival sets with one another (and by contrasting the members of that revival set with ideas where this distinguishing act has not (yet) been performed). Thus, non-distinguishability becomes the *definition* of "simple idea", that by which the members of its revival set are determined. To deploy this idea as the *explanans* of the systematic variety of human thought, though, more is needed. We must additionally take such simple ideas to play a certain *causal role*, and we do that by subsuming them under yet more general terms such as "images" or "building blocks". Consider the latter general term. Whereas we have no experience of seeing our complex ideas being constructed out of their simple components (in part because we have no experience of their simple components as such), we do have experiences of, for example, a house being constructed from bricks, or a house being constructed from wood, or a boat being constructed from wood. Such experiences form the revival set for terms such as "building block" and thereby come to be represented as resembling one another. As we have seen, it is by representing diverse experiences using a single general term, or representing them as being of a single kind, that Hume takes us to represent them as sharing certain causally efficacious properties. Thus, in subsuming the revival set of "simple idea" under the revival set of "building block" we represent simple ideas as sharing the causal powers of building blocks, in this case as being the elements from which something else is constructed. Complementarily, it is by subsuming these same ideas under the general term "images" or "perceptions" or "ideas" that we come to represent that they are more specifically the building blocks of our complex ideas. That is, it is by specifying the determinate ways that our theoretical posit (simple ideas) resembles (images, building blocks) and differs from (distinguishability) its perceptible model (complex ideas, wood and bricks) that we explain the manifest phenomena as being the result of the causal powers of some underlying substance.

Having done so, we check the reliability of this theoretical posit against experience to ensure that it accords with our best evidence, and succeeds in improving on the evidence offered by induction alone (c). In the case of simple ideas, Hume uses that theoretical posit to account for not only the general reliability of taking ideas to depend on their correspondent impressions, but also the otherwise puzzling cases of ideas that do not exhibit this dependence, e.g., the ideas of Paris, New Jerusalem, and more controversially the missing shade of blue.[21] The inductive evidence alone points to some connection between impressions and ideas, but it is only by leveraging the simple-complex distinction that we can represent the real regularities that these perceptions exhibit (that all simple ideas are copies of some simple impression). Especially in combination with the associations of ideas and the division of the productive faculties into memory, imagination, and reason, what that posit accomplishes is precisely a way to render any given complex idea expected. A path can always be traced from original impressions through the machinations of one or more cognitive faculty to the target idea, thereby rendering it the expected and explicable outcome of those processes. That this path can be traced from such a wide variety of cases is how the reliability of the theoretical posit is successfully checked against experience, which provides evidence for its truth.

The impression-idea distinction can be accounted for similarly, and so I will keep that explication brief. Reason discovers via induction certain universal (or near universal) regularities of experience: e.g., that the vulgar appear to reliably and systematically sort their perceptions into two distinct kinds, often but not always by their degree of force and vivacity (a). In this case, reason does not have a particularly large role in forming a theoretical representation of the substance that underlies this phenomenon (b) because the distinctions of the vulgar (between feeling and thinking) are already fairly close to marking the real distinction between the substances responsible for this regularity. What reason does do, though, is locate the *real* difference in the fact that impressions are mental originals whereas ideas are copies of these, thereby subsuming "impressions" and "ideas" under general terms such as "original" and "copy", respectively, that serve to represent the causal powers of the substances (impressions and ideas) posited (d-g). That is, it takes *this* difference to be the one that most reliably explains all the other differences at hand, and therefore the one to which we ought to ontologically commit ourselves by replacing the language of the vulgar ("feeling" and "thinking") and the language of the false philosophy (Locke's use of the "idea") with that of the scientist of human nature (j). The explanatory force underlying this commitment derives from the fact that it is because an impression *is* a mental original and an idea *is* a copy that, for example, the former tends to have a higher degree of force and vivacity than the latter. Thus, the manifest phenomena (these degrees of force and vivacity) is explained by the theoretical posit insofar as the theoretical posit renders these phenomena expected: the manifest phenomena are exactly what one would expect were the underlying substance as it is posited to be (i). Finally, this hypothesis is checked against experience and found to be accurate (c).[22]

Those are two examples of the process that I have argued governs Hume's pursuit of the science of human, and hopefully it is relatively clear how this same process could be used to account for other such theoretical posits. Rather than being a mere inductivist, a radical skeptic, or an extreme metaphysical realist, Hume is a scientific realist. He believes that the distinctions that he draws in conducting the science of human nature reflect the real causal powers of the substance underlying the manifest phenomena of the human mind. He assigns reason and the senses each a role in discovering the inductively-established universal regularities that demand explanation, in constructing representations of the theoretical entities posited to do that explaining, and in gathering the evidence of the truth of representations. As Hume sees it, his own pursuit of the science of human nature is a smashing success insofar as it explains many of the universal regularities governing the mind, uncovers the deep ontology of, does not succumb to the intemperate desire for explanation to which his predecessors fall prey, and both explains and explains away the successes and failures of the system of the vulgar and the false philosophy. There is, however, one area in which Hume reports that his pursuit of the science of human nature has failed: in its attempt, "to explain the principles, that unite our successive perceptions in our thought or consciousness" (T App.20; SBN 636). As we have seen in the cases of the treatment of the ideas of body and necessary connection by the vulgar and the false philosophy, it can be revelatory to study not only the cases in which a theory succeeds in its explanatory task, but also those in which it fails. What is illustrative in those two cases is equally so in the case of Hume's own failure, and it is to the details of that failure that I turn in the final chapter.

Notes

1. Strawson, *Secret Connexion*, 18.
2. There is, of course, a substantial secondary literature on how precisely we are to understand this mistake of the vulgar. Part of that literature consists of scholars balking at Hume's suggestion that the "determination of mind" that is supposed to be the origin of this confusion is not itself a perception, and is therefore unsuited to serve as the model for an idea of power, or necessity, etc. That is, if an idea roughly represents that of which it is a copy, then since a determination of the mind is not itself a perception, but a certain behavior of perceptions, then it is not something that can be copied or therefore represented. (See Wolff,

"Hume's Theory of Mental Activity," 298; Noonan, *Routledge Philosophy Guidebook to Hume on Knowledge*, 141; Bennett, *Learning from Six Philosophers*, 305; Allison, *Custom and Reason in Hume*, 187.) The emended version of the Representational Copy Principle with which I have been working combined with the notion of a perceptible model (which Hume explicitly cites in just this context—see the discussion of T 1.3.14.20; SBN 164–5 below) handle this objection easily enough. The determination of the mind at issue is the customary transition from one perception to another. This transition can be represented easily enough via the schema,

'x'R'y' represents xRy,

thusly,

'x' customarily precedes 'y' represents x customarily precedes y.

That is, we can represent the customary transition from one perception to another by forming ideas of each of these perceptions that are themselves connected by a customary transition. This complex idea in turn can serve as the model for a new complex idea that resembles it insofar as it represents two objects as being constantly conjoined with one another, but differs from it insofar as it represents this relation as holding not between our own ideas, but rather between the objects of those ideas (which, of course, the vulgar take to be certain distinct and continued existences, and the false philosophy takes to be matter).

Now, that way of casting the determination of the mind as the source of the idea of power, etc., appears to run into another stumbling block from the secondary literature: that Hume appears to understand both the determination of mind and the confused representation of power that it engenders to operate *pre-reflectively*. If, however, the representation of power operates pre-reflectively, then the interpretation above wherein this representation is the result of modeling it on an antecedently formed idea of the determination of mind itself, both that determination and the attribution of a similar relation to the objects of our ideas seem to be the result of deliberate and reflective action.

Notice that on the above interpretation what it is to form a representation of two objects (be they impressions, ideas, matter, whatever) as containing certain powers is just to form a certain association between the ideas of those objects. Thus, while it is *possible* to form the idea of power by explicitly modeling it on the idea of the determination of mind—which is what the first set of objections demands be proved—that is entirely compatible with our *actually* forming that idea merely by taking up that custom or association, which is what the second objection demands. Thus, the Representational Copy Principle as I understand it provides a single simple solution to both of these traditional thorny issues, especially when keeping in mind Hume's thesis that the vulgar idea of necessary connection is itself a massive confusion.

My thanks to Joshua Wood for his very helpful framing of these issues in his presentation at the *41st Annual Hume Society Conference*, "Hume, Causal Power, and Conceptual Empiricism," and for his subsequent generosity in sharing his notes for that presentation.

3. Strawson dubs the ambiguity in Hume's texts on the question of the whether he rejects certain ideas, such as those of necessary connection, the external world, and the self, as "unintelligible" or "meaningless" (and what he means by those phrases when he does employ them) the Meaning Tension (Strawson, *Secret*

Connexion, 118). He purports to resolve this tension by attributing the two-part account of representation to Hume that we examined in the previous chapter. Talk of the meaninglessness of certain ideas amounts to talk of their lack of descriptive content, whereas talk as if they are meaningful is made possible via the deployment of relative ideas, which have no descriptive content, but make bare reference nonetheless. Given the objections I raised to that view there, I believe that understanding Hume as employing a mechanism of whereby the confused, muddled, and ultimately meaningless language of the vulgar and false philosophers is explained and replaced by that of the scientist of human nature is our best hope of resolving this tension. In any case, Broughton's and Beebee's apparent blind spots for the tension itself are an unfortunate oversight.

4. Broughton, "Hume's Ideas about Necessary Connection," 225.
5. Broughton, "Hume's Ideas about Necessary Connection," 226–8.
6. Beebee, *Hume on Causation*, 117–18.
7. Recall that Hume notes that, "'tis very difficult to talk of the operations of the mind with perfect propriety and exactness; because common language has seldom made any very nice distinctions among them, but has generally call'd by the same term all such as nearly resemble each other" (T 1.3.8.15; SBN 105).
8. See also T 1.3.14.15; SBN 162.
9. Notice that Hume uses the contrived idiom of "containing" power or efficacy, perhaps because the Representational Copy Principle is at most agnostic as to whether impressions represent. So, it is awkward for Hume to write that we have no impression *of* power or efficacy. What he means to indicate is that no impression has a necessary connection as part of its pictorial character, i.e., that itself consists of a necessary connection between distinct existences.
10. See also 1.3.3.3; SBN 79–80.
11. As Sellars puts it, this makes the manifest image prior in the order of knowing, but posterior in the order of being. Sellars, "Scientific Realism or Irenic Instrumentalism," §57.
12. In the *Enquiry* Hume is more explicit that stable expectations are not only produced by casual reasoning, but are also its end. For example,

> In reality, all arguments from experience are founded on the similarity, which we discover among natural objects, and by which we are induced to expect effects similar to those, which we have found to follow from such objects. [. . .] From causes, which appear *similar*, we expect similar effects. This is the sum of all our experimental conclusions.
>
> EHU 4.20; SBN 36

The sum of all experimental conclusions is that we expect similar effects from similar causes. The conclusion of a bit of causal reasoning, that at which causal reasoning aims, is the expectation that that a manifest *explicandum* will occur, given the nature of its *explanans*. Hume is likewise explicit that this expectation is the result of specifically *substantial* causal explanation.

> But notwithstanding this ignorance of natural powers and principles, we always presume, when we see like sensible qualities, that they have like secret powers, and expect, that effects, similar to those which we have experienced, will follow from them.
>
> EHU 4.16; SBN 32–3

It is by explaining the observed regularities of experience by appealing to the natural powers and principles of the substance underlying such regularities that we come to form the (forceful and settled) expectation that like effects will follow from like causes. See also, EHU 4.21; SBN 36–7, EHU 5.8; SBN 46–7, EHU 6.4; SBN 57–9, EHU 7.21; SBN 69–71, EHU 9.1; SBN 104–5.

13. Again, Loeb has a similar take on this passage, noticing the importance that the long-term stability of beliefs plays here. Loeb, *Stability and Justification*, 60–5. And again the difference between us is that Loeb sees this importance as owing to the relief from "the felt uneasiness in unstable doxastic conditions" whereas I take it to be a reliable indicator of truth. Thus, I would emphasize the close pairing that Hume makes between stability and *explanation*, e.g., in the passages from the Introduction discussed in Chapter 2 (T Intro.9; SBN xvii-xviii).
14. This is not unrelated to Wolff's casting the association of cause and effect as a second-order disposition to develop a first-order disposition to expect a cause to be followed by its effect. (Wolff, "Hume's Theory of Mental Activity," 293–4.) These meta-expectations due to the general reliability of reason are the flipside of the diminishing degree of confidence in our beliefs that is the result of focusing our attention on reason's occasional fallibility, which Hume describes in 1.4.1, "Of scepticism with regard to reason." Thus, I follow Garrett in understanding 1.4.1 as presenting one among many possible ways that our *de facto* reflections on our own reasoning might go rather than giving a *de jure* negative epistemic evaluation of our beliefs. Garrett, *Cognition and Commitment*, 222–8.
15. Recall that the difference that Hume reports between his ideas of New Jerusalem and Paris is not that one is more detailed than another—the point of the latter example is that his idea of Paris is vague and inaccurate—but rather lies in the differences in the way that each relates to that which it represents. T 1.1.1.4; SBN 3.
16. De Pierris, "Hume and Locke on Scientific Methodology," 279.
17. See, for example, T 1.3.12.25; SBN 142, T 1.3.13.8; SBN 147, T 1.4.2.42; SBN 208–9, T 2.1.12.6; SBN 327, IS 31, EHU 9.1; SBN 104–5, EHU 11.30; SBN 147–8.
18. My thanks to Katharina Paxman and Margaret Schabas for prompting me to think more about Hume's understanding of analogy, and to Miren Boehm for specifically pointing me to the *Dialogues* as an important source on the topic.
19. Of course, it is only after having drawn the simple-complex distinction in this way that we are in a position to retrospectively describe this process as discovering the resemblance between the objects of the ideas called by the name "simple".
20. It is worth recalling that, as we have seen in our discussion of external existence and necessary connection, and as we will see in discussing personal identity, Hume is steadfast in holding to that theory of mental representation even in the face of the most extreme examples of such conclusions.
21. Hume takes the missing shade of blue to be an exception to our ability to so trace all of our ideas, but he also famously reports of that case that, "tho' the instance is so particular and singular, that 'tis scarce worth our observing, and does not merit that for it alone we should alter our general maxim" (T 1.1.1.10; SBN 6).
22. T 1.1.1.9–11; SBN 5–7.

6 Explanation and Personal Identity in the Appendix

The main thesis of this chapter will be that the interpretive line on Hume's understanding of the science of human nature that I have been defending brings a degree of clarity to Hume's dissatisfaction with his theory of personal identity in the Appendix. What I will argue is that the theory of personal identity that Hume presents in the body of the *Treatise*, like the accounts of external existence and necessary connection already considered, consists of an extended argument for the conclusion that we can have no idea of a single subject of experience persisting through time. Once again, Hume's argument takes the now familiar form of eliminating in turn the three kinds of ideas that I have claimed Hume finds us able to form: simple ideas, straightforward complex ideas, and theoretical representations. What Hume realizes in reconsidering that account between the publication of the *Treatise* and the writing of the Appendix is that while he continues to hold that that account provides an adequate *description* of our experience of personal identity, that description itself produces an explanatory demand that, given his elimination of the relevant possibilities for providing a *theoretical* account of personal identity, cannot be met. Thus, the contradiction that Hume reports discovering is not one that is internal to his theory of personal identity itself, but is rather that which arises from the demand for an explanation that that account presents and the impossibility of meeting that demand.

That is a very general description of the interpretive theses that I will defend in this chapter, but before beginning that work, it will be worth going through that outline again in a little more detail. Consider then, that in the Appendix, Hume famously cites two principles concerning personal identity that he takes to be inconsistent, but neither of which he can abjure:

(a) "that all our distinct perceptions are distinct existences," and
(b) "that the mind never perceives any real connexion among distinct existences."

T App.21; SBN 636

The most striking thing about these principles is that they are not, in fact, inconsistent.[1] Scholars working on this problem have presented a wide variety of suggestions of what Hume might be thinking here.[2] I will argue that such scholars have largely overlooked an important clue: the fact that Hume presents these two principles specifically as a way of expressing his dissatisfaction with any theory that purports, "to explain the principles, that unite our successive perceptions in our thought or consciousness" (App.21; SBN 635). The key here, I will suggest, lies in understanding Hume's notion of explanation.[3] In fact, Hume himself frames his worry about his theory of personal identity in terms of its explanatory power not once, but twice in the paragraph in which he first introduces his concern.

> But having thus loosen'd all our particular perceptions, when I proceed to *explain* the principle of connexion, which binds them together, and makes us attribute to them a real simplicity and identity; I am sensible, that my account is very defective [. . .] all my hopes vanish, when I come to *explain* the principles, that unite our successive perceptions in our thought or consciousness. I cannot discover any theory, which gives me satisfaction on this head.
>
> T App.20; SBN 635–636, emphasis mine

What (a) and (b) combine to demonstrate is that *if* there is a real connexion among our distinct perceptions, it is not one that we perceive. Thus, any theory that purports to explain the principles that unite our successive perceptions, must be one that makes an appeal to unobservables. As I have argued, while there was a time in Hume scholarship when this would have been enough evidence to credit Hume with rejecting any such theory, there is good reason to suppose that explaining some manifest phenomena via an appeal to unobservables is not only acceptable, but also an essential part of Hume's pursuit of the science of human nature. If that argument has been even moderately successful, though, then Hume's rejection of theories purporting to explain the unity of consciousness cannot be based solely on the deployment by such theories of unobservables.

Rather, I will argue, the inconsistency is generated from the combination of the demand that the principles of the unity of consciousness be explained and the impossibility of providing such an explanation. While Hume himself might well posit certain unobservable entities in his explanations of human nature, one of the conditions for his legitimately doing so is that he be able to account for our ability to *represent* such entities even without their being observed. As we have seen, the form that this account takes in Hume is providing a perceptible *model* for such entities and specifying the ways that the posited unobservables resemble and differ from this model. The problem with the demand to explain "the principles, that unite our successive perceptions" is that there is *no* model available for the entities that they must posit in order to explain that unity, and so there is no possible theory that allows us to represent the unity of consciousness.[4] As Hume concludes, we must therefore accept, "with regard to the mind, that we have no notion of it, distinct from the particular perceptions" (T App.19; SBN 635), i.e., that we cannot move beyond what is observable, in this case "the particular perceptions," to a more robust theoretical-explanatory account.

The contradiction that Hume confronts in the Appendix arises from the combination of this failure with the ineliminable demand for an explanation that the principles of connection among our perceptions nonetheless place on the science of human nature.[5] The key here is to see that it is this principle of connection itself that is the intended *explicandum* of a theory of personal identity, but Hume's account in the body of the *Treatise* amounts to nothing more than a description of that very principle, and so cannot also serve as its explanation. What Hume discovers are the regularities that govern the behavior of our perceptions: one perception follows another when the two are connected in a certain way via relations of resemblance and causation. As we have seen, though, for Hume merely being able to subsume some phenomenon under an empirical generalization like this one is not to *explain* that phenomenon. What needs to be explained is precisely the fact that the particulars (here, perceptions) exhibit such regularities (here, the principles of connection).

My procedure will be to explicate each side of Hume's "contradiction" in turn. The first section will address the impossibility of producing a representational model for the unity of consciousness that Hume discovers in 1.4.5 and 1.4.6.[6] I will there use the account of Hume's understanding of the science of human nature that I have been developing to recast the issue of Hume's concern with personal identity from one way of conceiving it—as a problem with unobservable entities in general—to its proper place as a problem with the intelligibility of any possible *representational modeling* that would be adequate to the task of explaining the unity of consciousness. Then, I will address the ineliminable demand for an explanation of the principles of connection among our perceptions in the context of Hume's pursuit of a science of human nature by returning to the Appendix to demonstrate how this failure can be used to account for the contradiction that Hume cites there as presenting an insurmountable obstacle to the success of his account of personal identity.

The Impossibility of a Representational Model for Personal Identity

As I have been arguing, Hume does not endorse a blanket prohibition on the use of theoretical entities or unobservables, and in fact makes use of them in precisely the way that one would expect: as explanatory posits necessary for the project of pursuing the science of human nature. Such posits are legitimate just in case they explain some phenomenon using a perceptible model. More specifically, I have argued that, *contra* one line of interpretation, rather than proceed by subsuming particulars under inductive-established universal regularities of experience, scientific explanations consists of explaining such regularities themselves by appealing to the nature, powers, or essence of the substance that underlies such manifest phenomena. The two most prominent examples of this kind of explanation that I have presented are Hume's use of the impression-idea and simple-complex distinctions, although we have also encountered other examples along the way (some provisional, e.g., the imperceptible air or third spatial dimension, and some more steady, e.g., the positing of reason as a distinct cognitive faculty). As Hume notes in the Appendix, in the case of personal identity, the *explicandum* at hand is, "the principle of connexion, which binds [our particular perceptions] together." That is, it is possible to discover certain inductively-established universal regularities governing the human mind, most importantly those that prompt us to think of each human mind as a kind of unity. Hume identifies these "principles of connexion" as resemblance and cause and effect (although he cites only the latter as constituting the "true idea of the mind"). The unity that such principles describe is the *explicandum* of a theory of personal identity, and the demand to explain this unity is most naturally understood, and had been understood by Hume's predecessors, as a demand to provide an account of that single thing which underlies those perceptions and in which they inhere or of which they are parts (T 1.4.5.2; SBN 233). That is, as Hume understands it, prior to the *Treatise*, accounts of personal identity had aimed at formulating theoretical posits intended to explain the regularities governing the human mind. Again, Hume has no objections to that mode of explanation *per se*, and in fact regularly employs it himself, but his own theory of personal identity is meant to show that no such account is possible in this case.

What a successful explanation of the principle of connection that binds our perceptions together would require is a model, taken from a purely descriptive phenomenology, and a specification of how the theoretical entity posited both resembles and differs from that model. Recall one such case that we considered earlier: the Bohr model of the atom. Taking the solar system as its perceptible model, Bohr proposes that we understand the structure internal to an atom as analogously consisting of a large nucleus being orbited by smaller electrons, but also specifies determinate differences between the theoretical entity posited and its perceptible model: the former is orders of magnitude smaller, it is held together by electrostatic forces instead of gravity, the electrons' orbits are circular rather than elliptical, etc. Employing such a model accomplishes two tasks. Most relevant at the moment is that it allows us to expand our representational capacities from that which is "directly" observable, in this case the solar system, to that which is not, the internal structure of an atom. Going forward, it will also be important to note that, if successful, such a model also allows us to explain the observed regular behavior of its *explicandum* via an appeal to its posited unobservable nature. In this case Bohr's model would allow us to explain the Rydberg formula for the spectral emission lines of atomic hydrogen via the posited internal structure of an atom.

Hume begins constructing his argument that no such model is possible in the case of personal identity in the section on the immateriality of the soul, which raises the same problem for this kind of substance as was previously raised regarding material substance, and more still.

> In order to put a stop to these endless cavils on both sides, I know no better method, than to ask these philosophers in a few words, *What they mean by substance and inhesion?* [. . .] As every idea is deriv'd from a precedent impression, had we any idea of the substances of our minds, we must also have an impression of it; which is very difficult, if not impossible, to be conceiv'd. For how can an impression represent a substance, otherwise than by resembling it? And how can an impression resemble a substance, since, according to this philosophy, it is not a substance, and has none of the peculiar qualities or characteristics of a substance?
>
> T 1.4.5.2–3; SBN 232–233

As we did in the previous two chapters, it is worth pausing here to note that once again Hume's conclusion is explicitly and repeatedly concerned with the nature of our *representation* of ourselves. He asks the proponents of the immaterial soul what they *mean* by 'substance', and how an impression can *represent* the substance of our mind, etc.[7] Both in 1.4.5 and in 1.4.6, Hume is concerned with the limits of our representational capacities, and as we have noted previously, combining this concern with Hume's theory of mental representation gives Hume's arguments a very specific structure. To conclude that we have no idea of a self other than that of a bundle of perceptions, Hume will have to show that we can have no such simple idea, no such straightforward complex idea, and no theoretical representation of any such thing. In this case, though, while Hume does give arguments in passing regarding the simple idea of a soul-substance, his focus from the start is on the purported theoretical representation of it instead.

Returning to the passage above, we can see that Hume's argument there already engages the question of whether the soul can be represented by a theoretical posit. Neither impressions nor ideas can represent mental substance because they do not resemble it in the right way. Specifically, they have "*none* of the peculiar qualities or characteristics of a substance." If an impression had *some* of these qualities or characteristics, it could provide a perceptible model for mental substance. For example, if mental substance was purported to be a perception with even more force and vivacity than impressions, or of a color that no impression could replicate, we could at least get a handle on what such a substance would be like. We could specify determinate ways that such a substance resembled other perceptions and differed from them, and in that way could generate a representation of it.

That mental substance is supposed to share *none* of the qualities or characteristics of perceptions, however, means that without direct perceptual access to it, we cannot even so much as represent it via such a model.

After challenging the proponent of the representation of the soul to give an account of the place of such an idea in our cognitive apparatus, Hume considers yet another proposal for representing the soul via a theoretical posit. Specifically, he considers whether the soul can be represented as "*something which may exist by itself*," a phrase to which he repeatedly refers as a *definition*. As we have seen previously, for Hume to give a definition is to specify the revival set for a given general term, which general term represents its members as resembling each other.[8] Thus, while the proponents of the idea of a soul might intend this definition to "reach beyond" our perceptions to refer to that in which such perceptions "inhere"—notice the similarity between this theory of reference and that deployed by Strawson to garner the appropriate reference for his interpretation of "relative ideas"—Hume will have none of that. As I have argued, while we can use general terms to represent some distinctions as corresponding to real relations among perceptions (by adopting some set of general terms and abandoning others), that procedure itself requires being able to delimit determinately the members of the relevant Revival Sets (by specifying the ways that represented theoretical entities resemble and differ from their perceptible models). We have previously observed two ways of failing to meet these conditions: a potential theoretical definition can fail to fix a determinate revival set because the entities that it posits do not *resemble* any perceptible model in any way, or because the entities it posits do not *differ* from any perceptible model in any way. Hume first casts the attempted posit of the proponents of the soul as failing in this second way.

> My conclusion [. . .] is, that since all our perceptions are different from each other, and from every thing else in the universe, they are also distinct and separable, and may be consider'd as separately existent, and may exist separately, and have no need of any thing else to support their existence. They are, therefore, substances, as far as this definition explains a substance.
>
> T 1.4.5.5; SBN 233

The definition that the proponent of the idea of a soul gives is that a soul-substance is "something which may exist by itself." Finding that all perceptions may exist by themselves, Hume concludes that this definition is empty: it does not specify any determinate revival set because the perceptions that it ends up including in its scope are *all* perceptions. A moment later, Hume puts this same conclusion a slightly different, but complementary way.

> We have no perfect idea of any thing but of a perception. A substance is entirely different from a perception. We have, therefore, no idea

> of a substance. Inhesion in something is suppos'd to be requisite to support the existence of our perceptions. Nothing appears requisite to support the existence of a perception. We have, therefore, no idea of inhesion.
>
> T 1.4.5.6; SBN 234

The idea of a soul-substance is the idea of that which can exist by itself. That idea encompasses all perceptions. Its complement is the idea of inhesion, or dependence. It is the idea of that which requires something else for its existence. Whereas the former idea includes all perceptions, the latter idea includes none of them, and so it too is an idea that can have no content. The former specifies no *difference* between that which it purports to posit and that posit's model; the latter specifies no *resemblance* between these, and thus neither has any content. We have no idea *of* either inhesion or soul-substance.

As Hume points out at the start of 1.4.5, since the attempt to posit a soul-substance is an instance of the more general mistake of misunderstanding substance as a nothing more than a *via negativa*, the refutation of that posit is more or less an application of the same reasons that supported the rejection of the idea of external existence in 1.2.6 and 1.4.2.[9] Accordingly, just as the false philosopher's attempt to posit matter as that which explains the apparent constancy and coherence of our perceptions is complemented by the vulgar's attempt to posit the existence of distinct and continued *perceptions*, here the attempt to posit a soul-substance is complemented at the start of 1.4.6 "Of personal identity" with an attempt to posit a *perception* that explains person's identity across time. As was the case with the idea of body, Hume appears to take this posit more seriously than he does that of the false philosophers, but does also find it easily refuted. For example, just as he gave an easily obtained, but methodologically sound empirical refutation of the continued and distinct existence of certain perceptions (the experiment of pressing on one's own eye), so too does he begin his refutation of the thesis that, "we are every moment intimately conscious of what we call our SELF."

> Unluckily all these positive assertions are contrary to that very experience which is pleaded for them, nor have we any idea of *self*, after the manner it is here explain'd. [. . .] If any impression gives rise to the idea of self, that impression must continue invariably the same, thro' the whole course of our lives; since self is suppos'd to exist after that manner. But there is no impression constant and invariable. Pain and pleasure, grief and joy, passions and sensations succeed each other, and never all exist at the same time. It cannot, therefore, be from any of these impressions, or from any other, that the idea of self is deriv'd; and consequently there is no such idea.
>
> T 1.4.6.2; SBN 251

The second sentence here makes clear that once again Hume plans to leverage the Representational Copy Principle in order to conclude that, as he puts it in the first sentence, we have no "idea *of* self, after the manner it is here explain'd." It is because an idea represents (roughly) that of which it is a copy that if we are to have an idea of the single subject of experience persisting through time that we must have a single impression that persists through time. Unlike in the case of the idea of a soul-substance, though, here Hume's argument begins not with the claim that no such idea is *possible*, or that there is anything misguided in the very pursuit of such an idea, but rather with the simple observation that we, in fact, have no such idea. It is *experience* that shows that rather than have a single impression that remains constant and invariable throughout our lives, our perceptions are diverse and constantly changing.[10]

Of course, as the portion of the above passage that I omitted above makes clear, as was true in the case of the vulgar idea of body, so here too Hume's ultimate target is the very intelligibility of the idea of a distinct and continued self, be it conceived of as a soul-substance or a perception unlike any other.

> For from what impression cou'ld this idea be deriv'd? This question 'tis impossible to answer without a manifest contradiction and absurdity and yet 'tis a question, which must necessarily be answer'd, if we wou'd have the idea of self pass for clear and intelligible. It must be some one impression, that gives rise to every real idea. But self or person is not any one impression, but that to which our several impressions and ideas are suppos'd to have a reference.
>
> T 1.4.6.2; SBN 251

What is at stake in this section is whether the idea of self can pass for clear and intelligible, and Hume's conclusion is that it cannot. We cannot cite the impression from which such an idea would be copied without manifest contradiction and absurdity. As above, Hume's intention is to show that we have no idea of the self as conceived in this way. Of course, as we have seen repeatedly, Hume's claim here that, "It must be some one impression that gives rise every real idea," is a bit of hyperbole. While every *simple* idea must be derived from some one impression, straightforward complex ideas and theoretical representations do not have to be so derived, and so Hume's argumentative burden will once again not be as easy as he at first portrays it.[11] That said, the claim that follows this one goes a long way towards making that more substantial case: "But self or person is not any one impression, but that to which our several impressions are suppos'd to have a reference." Here again is the suggestion that we can refer to that which we cannot conceive. It is striking that Hume does not even so much as pause to refute this suggestion here, but rather follows its presentation with the instantiation of the Representational Copy Principle above. What that sequence implies is that the Representational Copy Principle accounts for all of the ideas that we can have—relative or otherwise—and that no idea can have a reference to anything outside of its scope. If non-descriptive reference (a la Strawson) were a live option for Hume here, he would surely give it more attention when presenting this argument.

If the Representational Copy Principle is supposed to eliminate the possibility of an idea of the self qua *subject* of experience, though, how precisely does it do so? Here is how I see that argument going. Modeling the self on perceptions themselves will not work because the key difference between the self and perceptions is one to which we can give no independent content: it requires an entirely radical departure from experience. In order to articulate this difference, we would need to posit a perception that again, "has none of the peculiar qualities or characteristics of," other perceptions. We are asked to represent a perception that is not the object of any particular perception, but its "subject" and there is no way to articulate this difference using just the resources that Hume allows. For example, while reason and the imagination are adept at forming complex ideas via the rearrangement of simple ideas, since the idea of the subject of experience is one that is supposed to transcend the content of any idea whatsoever, no such rearrangement can be adequate to its representation.

Likewise, even moving beyond the standard associative mechanisms of contiguity, resemblance, and cause and effect would not be of any use, as again the problem is that these relations are used to structure our complex ideas, and merely changing this structure would still leave the problem that the content of any idea of the subject of experience is supposed to be radically different from the content of any other perception.[12] That which is the subject of a perception rather than the perception itself or its object does not differ from these in some determinately specifiable way, but would have to be of an entirely different kind. As Garrett puts it in discussing not just the idea of the self, but also any other purported counterexample to Hume's Copy Principle,

> In each of these cases, admitting a counterexample to the Copy Principle would mean not merely violating the Resemblance Thesis but violating it in such a way as to allow nonimagistic ideas that could not, even in principle, resemble impressions. It would thus require the admission of an entirely distinct representational faculty, and hence a very serious modification in the cognitive psychology that Hume thinks he finds otherwise well supported by experience.[13]

While the line that I have been pursuing here gives more leeway to the scope of our representational faculties than employing merely the Copy Principle to that end would, Garrett's more general point stands. Hume's predecessors had supposed that we could represent certain contents that, according to Hume's account of our representational capacities, are strictly speaking impossible to represent. The self, qua either mental substance or a single perception that is itself the subject of all representation, is one such idea. We neither perceive such self, nor can construct a perceptible model of it, and so the idea of the self cannot "pass for clear and intelligible" (T 1.4.6.2; SBN 251).[14]

That the account of personal identity under consideration in the opening of 1.4.6 is one that attempts to form a specifically theoretical representation of the self is made clear by Hume's discussion of whether the self persists when we the perceptions of which it is the purported subject do not. That Hume takes this question up indicates that the view that he takes himself to be refuting is one that posits an unobservable theoretical-explanatory entity, and Hume's conclusion is that no such posit is in fact possible.

> When my perceptions are remov'd for any time, as by sound sleep; so long am I insensible of *myself*, and may be truly said not to exist. And were all my perceptions remov'd by death, and cou'd I neither think, nor feel, nor see, nor love, nor hate after the dissolution of my body, I shou'd be entirely annihilated, nor do I conceive what is farther requisite to make me a perfect non-entity.
>
> T 1.4.6.3; SBN 252

If the line that I have been defending is correct, then Hume's strong and radical conclusion here—that we do not exist when we are soundly asleep—cannot be supported by a rejection of theoretical entities *simpliciter*, but must rather follow from the considerations specific to representing the self, and I believe that it does. At the very least, that is an interpretive line that is possible at this point in the dialectic that Hume has pursued since this conclusion comes on the heels of his rejecting what he takes to be the only two theoretical-explanatory posits available. That is, after 1.4.5 and the arguments from the opening of 1.4.6, Hume has concluded that neither the positing of a subject-perception nor that of a subject-non-perception will be able to have the representational content necessary to serve as an explanation of the unity of consciousness. With the failure of those two attempts, though, the possibility of a theoretical explanation of that unity dwindles, and so Hume concludes, as the methodological principles we saw in Chapter 2 prescribe, that no such explanation is possible here, and considers whether we must take the inductively-established universal regularities at hand (the principles of connection amongst our perceptions) as themselves explanatorily basic. It is to an investigation of these regularities, which seem to be precisely the kind of phenomena that the science of human nature sets out to explain, that he turns his attention.

> What then gives us so great a propension to ascribe an identity to these successive perceptions, and to suppose ourselves possest of an invariable and uninterrupted existence thro' the whole course of our lives?
>
> T 1.4.6.5; SBN 253

Given the close parallels that we have already observed between Hume's account of the idea of body and that of personal identity, it is unsurprising to find that Hume attributes the confusion he finds in the latter to a mistake similar to the one at the root of the former.

> That action of the imagination, by which we consider the uninterrupted and invariable object, and that by which we reflect on the succession of related objects, are almost the same to the feeling, nor is there much more effort of thought requir'd in the latter case than in the former. The relation facilitates the transition of the mind from one object to another, and renders its passage as smooth as if it contemplated one continu'd object. This resemblance is the cause of the confusion and mistake, and makes us substitute the notion of identity, instead of that of related objects. However at one instant we may consider the related succession as variable or interrupted, we are sure the next to ascribe to it a perfect identity, and regard it as invariable and uninterrupted. Our propensity to this mistake is so great from the resemblance above-mention'd, that we fall into it before we are aware; and tho' we incessantly correct ourselves by reflection, and return to a more accurate method of thinking, yet we cannot long sustain our philosophy, or take off this biass from the imagination.
>
> T 1.4.6.6; SBN 253–4

Once again we find Hume sighting the *bias* of the imagination to run together ideas for the sole reason that they resemble one another as the source of a serious cognitive error. It is significant, though, that just as in the case of the idea of body, this error does not *constitute* the mistaken idea at hand, but rather provides illegitimate *grounds* for what would otherwise be legitimate theoretical activity. Hume continues,

> In order to justify to ourselves this absurdity, we often feign some new and unintelligible principle, that connects the objects together, and prevents their interruption or variation. Thus we feign the continu'd existence of the perceptions of our senses, to remove the interruption; and run into the notion of a *soul*, and *self*, and *substance*, to disguise the variation.
>
> T 1.4.6.6: SBN 254

The unintelligible posit of a soul, and self, and substance is a theoretical posit intended to *justify* the absurdity of running together that which is identical with that which is merely resembling. That is, as in the case of body, the imagination, because of its biases, makes it appear as though we are confronted with an inductively-established universal regularity. As I have been arguing, such regularities, rather than being the *explanans* of scientific explanation, are its *explicanda*. Thus, the imagination's illicit activity makes it appear as though a theoretical explanation is called for when it is not. That, in turn, prompts us to attempt to construct a theoretical representation to do this explanatory work, and thus to speculate about a soul, or self, or substance that persists through these changes.

After a detailed examination into the fiction of identity that is the product of the imagination's biases, Hume returns to the "true idea of the human mind".

> As to *causation*; we may observe, that the true idea of the human mind, is to consider it as a system of different perceptions or different existences, which are link'd together by the relation of cause and effect, and mutually produce, destroy, influence, and modify each other. Our impressions give rise to their correspondent ideas; and these ideas in their turn produce other impressions.
>
> T 1.4.6.19; SBN 261

The representation of the human mind as being constituted by its causal structure is the *true* idea of it because, as we have repeatedly seen, representations of the causal structure of the world are the representations that Hume takes to be true ones. Such representations most reliably track the evidence afforded by experience, which is evidence of their truth, and insofar as they are reliable and therefore likely to be true, they represent the nature, powers, essence, and substance of the world.

Of course, as we observed in Chapter 1, the perceptions that stand in causal relations to one another, and thus constitute the substance of the human mind are not themselves perceptible as such. While our phenomenology is always complex, what explains the systematic connectedness of this complexity is that it is composed of *simple* ideas, which are themselves not perceived, but which are modeled on our complex ideas and posited on these explanatory grounds. Likewise, we do not perceive an ontological distinction between impressions and ideas, but rather posit that the perceptions that we sort into the revival sets of the terms "impression" and "idea" really are different and have differing causal powers. The distinctions between memory, imagination, and reason likewise mark differences in the ontology of the mind that are not directly observable, but are grounded in their explanatory force. What these previous conclusions imply for Hume's theory of personal identity is that the true idea of the human mind is likewise not merely an idea of some manifest phenomena, but rather the idea of the causal relations holding among these theoretical-explanatory entities.

Despite the fact that Hume's account of personal identity casts the human mind as consisting of certain theoretical entities standing in certain causal relations to one another, it is important to note the sense in which that account still differs from the previous accounts of the human mind as a soul-substance or as a perception with a continued and distinct existence. Specifically, in taking the human mind to be a bundle of perceptions united by relations of cause and effect, even though these perceptions themselves are theoretical entities, Hume's theory of personal identity does not *itself* make any new theoretical-explanatory posit. It remains, in an important sense, merely descriptive. Whereas, for example, the simple-complex distinction aims to *explain* the variety and systematic connectedness of human thought by appealing to its being constituted by simple perceptions, Hume's account of personal identity does no such thing. That account merely describes the systematic connectedness of perceptions more generally. The true idea of the human mind describes the constant conjunctions between our perceptions, but offers no explanation for why such constant conjunctions occur. Unlike the explanations that Hume offers elsewhere, here his account makes no appeal to the nature, powers, or essence of the substance underlying the inductively-established universal regularities of perception. That *explicandum* is described, but not explained.

Consider again Hume's example of the substantial properties of gold. Gold's color, its malleability, it's solubility in *aqua regia*, etc., are represented as substantial properties of gold insofar as we take these properties to themselves be attributable to the causal powers of the substance that underlies them. We take these properties to be substantial properties insofar as we take them to be the result of the causal powers of the substance gold. (We represent this substance, in turn, by adopting the term "gold", etc.) We explain the regular co-occurrence of these properties by supposing that there is some causal explanation for that co-occurrence. But now contrast this case with that of personal identity. In the latter case the causal relations at hand are those *between the substantial properties or perceptions themselves*. In the case of gold, we explicitly ruled this out as a way of understanding substantial explanation. It is not that we take gold's color to cause its malleability, or is solubility to cause its conductivity. Rather, the causal relation that these properties stand in is to their underlying substance. In the case of personal identity, though, it is precisely this kind of explanation that Hume has eliminated as a possibility. To what might we attribute the systematic connectedness of our perceptions? Hume considers two answers to that question—a soul-substance in which our perceptions inhere and a single perception that persists throughout our lives—and determines that neither is tenable. Thus, he turns to locating the true idea of the human mind in the causal relations *between* its perceptions. While that is, *prima facie*, precisely what Hume's own methodology prescribes that he does, it is important to note that in doing so, this becomes a case in which we abandon hope for an explanation. The causal relations that structure the mind, the constant conjunction of my impressions, their correspondent ideas, and the impressions and ideas that follow these must be taken as a brute fact. It is not attributable to the nature, powers, or essence of any underlying substance. A gas obeying the Ideal Gas Law is explained by casting it as a collection of molecules obeying familiar mechanical laws. The systematic connectedness of the content of our complex impressions given the relative paucity of experience is explained by casting complex ideas as arrangements of simple components. The point here is that there is no such explanation available for the mind. The mind's perceptions exhibit certain inductively-established universal regularities, for which there is not further account available.

That makes this a case in which Hume's own methodological imperative would dictate that he "sit down contented; tho' we be perfectly satisfy'd in the main of our ignorance, and perceive that we can give no reason for our most general and refin'd principles, beside our experience of their reality" (T Intro. 9; SBN xviii). As we are about to see, though, the problem with this prescription is that we cannot, in fact, sit down contented, perfectly satisfied with our ignorance because it is not *ignorance* alone that is at work in the case of personal identity. On the one hand, unlike the case of the idea of body, where the apparent universal regularities were explained *away* via an appeal to the biases of the imagination, the account of personal identity that Hume offers in 1.4.6 reveals that there really are such regularities governing the behavior of our perceptions more generally. Thus, the demand for an explanation in this case cannot be so easily ignored. On the other hand, as in the case of the idea of body, Hume's investigations into explaining these regularities via a theoretical representation of the substance underlying them have eliminated all such possibilities. So, it is not merely that we cannot *know* what explains these regularities, but rather that we cannot even so much as *think* what does. While it is one thing to rest content with ignorance so long as the possibility of an existent but unknowable explanation remains open, in this case what confronts Hume is the impossibility of any such explanation. Since, however, scientific practice is itself predicated on the existence of such explanations, what Hume discovers in the Appendix is that he stands in a kind of *methodological* contradiction. The inductively-established universal regularities that describe the unity of consciousness demand (substantial) explanation, but the results of his pursuit of the science of human nature yield the conclusion that no such explanation is possible.

The Ineliminable Demand for Explanation

What we have just seen is that no explanation of the unity of consciousness is possible. By itself, that is not enough to generate the contradiction that Hume reports finding in his account of personal identity. As I have suggested, that contradiction arises not within the account itself, but rather because of the combination the failure of that account to provide an *explanation* of the principle of connection of our perceptions and the fact that that principle nonetheless requires an explanation. Here we must tread carefully because paying close attention to what it is that stands in need of explanation is precisely what will reveal why it is that Hume finds his own account inadequate. Specifically, in the Appendix Hume repeatedly expresses his frustration at being able to explain *the principle of connection* among our perceptions. That this is what Hume takes the *explicandum* to be is of crucial significance because the account of personal identity that Hume offers in the body of the *Treatise* is precisely one that merely *describes* what this principle is, what the regularities are that govern the behavior of our perceptions. Because Hume holds that a theoretical-explanatory account of personal identity is not possible, he relegates his own account to the realm of descriptive phenomenology. In discovering *what* the principle of connection is among our perceptions, however, Hume (if successful) does no more and no less than describes that which the science of human nature demands be explained.[15] That description itself therefore presents Hume with the question of what perceptions *are* such that they are always found to be in accordance with these empirical generalities? This question is essential for Hume to answer precisely because it is the form of question that underlies the use of the kinds of explanatory models that are at the core of theoretical science, but which we have just observed Hume reject in the case of what is perhaps the most fundamental *explicandum* of a science of human nature: the human mind itself. That having a clear and accurate description of the phenomenon is a propaedeutic to explanation, but not itself an explanation, is what I propose Hume realizes between the publication of Book I and the addition of the Appendix, and is what makes him despair and retreat to skepticism in the latter.

To show that this it is this failure of explanation that is at the core of Hume's recantation, we can work through the relevant passages of the Appendix, noting how this hypothesis explains each claim that Hume there makes. Hume begins by restating his satisfaction with the accuracy of his own descriptive phenomenological account.

> This must pave the way for a like principle with regard to the mind, *that we have no notion of it, distinct from the particular perceptions*. So far I seem to be attended with sufficient evidence.
>
> T App.19–20; SBN 635

Next, he raises for the first time the concern that he has with this account. Notice that what he explicitly cites as standing in need of explanation is not the unity of our perceptions, but rather the *principle of connection* of these perceptions.

> But having thus loosen'd all our particular perceptions, when I proceed to explain the *principle of connexion*, which binds them together, and makes us attribute to them a real simplicity and identity; I am sensible, that my account is very defective, and that nothing but the seeming evidence of the precedent reasonings cou'd have induc'd me to receive it.
>
> T App.20; SBN 635, emphasis mine

It is the principle of connection which binds our perceptions together that is the intended *explicandum* of a theory of personal identity, but Hume's account amounts to nothing more than a description of that very principle, and so cannot also serve as its explanation. What Hume discovers are the *regularities* that govern the behavior of our perceptions: the perceptions that constitute the human mind all resemble one another (T 1.4.6.18; SBN 260–1) and "are link'd together by the relation of cause and effect" (T 1.4.6.9; SBN 261). As we must understand the latter relation according to Hume's earlier investigations—as consisting in these perceptions being of kinds that are merely constantly conjoined—what these discoveries amount to is simply that our perceptions are subject to certain empirically discovered generalities. As we have noted previously, though, merely being able to subsume some phenomenon under an empirical generalization like this one is not to *explain* that phenomenon. Hume is no mere inductivist. What needs to be explained is precisely the fact that the particulars (here, perceptions) exhibit such regularities (here, the principles of connection). What is called for in Hume's case, then, is not just to describe the principle of connection that governs our perceptions, but also to explain why it is that perceptions are subject to such a principle. What he sees as the *explicandum* here are the regularities that he himself describes our phenomenology as exhibiting.

Notice also that Hume reports having been mistakenly convinced by the "seeming evidence" of his precedent reasoning. Hume does not call into question the accuracy of his own account, but rather its evidentiary force.[16] On the interpretation I am defending, this fact is readily explicable: Hume's account is not evidence of anything because it is merely descriptive. A description of a certain phenomenon can serve as evidence only if it is evidence of some underlying ground for that phenomenon. Since Hume argues that there can be no genuinely explanatory account of personal identity (because we cannot form a perceptible model for the self), he thereby necessarily limits his account to a descriptive phenomenology, and thus circumscribed, such an account, no matter how accurate or informative it is, can never provide *evidence* for an underlying explanation.

Hume continues by once again presenting the empirical generalizations that he now finds inadequate.

> If perceptions are distinct existences, they form a whole only by being connected together. But no connexions among distinct existences are ever discoverable by human understanding. We only *feel* a connexion or determination of the thought, to pass from one object to another. It follows, therefore, that the thought alone finds personal identity, when reflecting on the train of past perceptions, that compose a mind, the ideas of them are felt to be connected together, and naturally introduce each other.
>
> T App.20; SBN 635

After declaring that his account cannot explain the connections between our perceptions, Hume recounts why it is that he was of necessity lead to that account in the first place. Perceptions are distinct existences and so the unity of consciousness results from their being connected in some way. These connections are not ones that we can observe, but they are ones that we can feel. Thus, to provide a descriptive phenomenology of these connections, one must make use of reflection, and it is this reflection that teaches us that the principles of the connection of our perceptions are felt ones: resemblance and cause and effect.

Hume then notes that as a description of the principles that constitute personal identity, this account is promising, but that when he attempts to *explain* these principles, he left at a loss.

> However extraordinary this conclusion may seem, it need not surprize us. Most philosophers seem inclin'd to think, that personal identity *arises* from consciousness; and consciousness is nothing but a reflected thought or perception. The present philosophy, therefore, has so far a promising aspect. But all my hopes vanish, when I come to *explain the principles*, that unite our successive perceptions in our thought or consciousness. I cannot discover any theory, which gives me satisfaction on this head.
>
> T App.20; SBN 635–636, emphasis mine

Again what we find here is Hume's hopes vanishing precisely when he comes to "explain the principles" that unite our perceptions. His concern is not that he has *misdescribed* those principles—causation and resemblance—but rather that these principles themselves stand in need of explanation, and that given the impossibility of providing a perceptible model for the self, the prospects for such an explanation are hopeless. Substantial explanation operates by explaining the regularities of experience via an appeal to the nature, powers, or essence of the substance underlying those phenomena. It is because our phenomenology *is composed of* simple ideas that we can have the complex ideas that we do. It is because impressions *are* mental originals whereas ideas are copies that the former have a higher degree of force and vivacity that they do. Because, however, Hume's investigations have yielded the result that there is nothing that the self *is* that the connections that we discover among our perceptions cannot be explained.

Thus, we come to Hume's declaration of the inconsistency that he finds in his account.

> In short there are two principles, which I cannot render consistent; nor is it in my power to renounce either of them, viz. *that all our distinct perceptions are distinct existences*, and *that the mind never perceives any real connexion among distinct existences*.
>
> T App.21; SBN 636

Our mind is composed of distinct existences, and we cannot perceive the "real" connection between these. What we can perceive, upon reflection, are the regularities to which these perceptions are subject—those of causation and resemblance—but it is our perceptions' being subject to such regularities that demands explanation, which explanation would have to appeal to the "real" connections that underlie them. Being prevented, as we have seen, from either perceiving or otherwise representing such connections, we are left in need on explanation that can never, in principle, be supplied.

Finally, Hume notes that were we capable of positing a substance in which our perceptions inhere, we could attain the needed explanation (in the same way that supposing that our complex perceptions *are composed of* simple ones explains the variety and systematic connection of our thoughts, or that supposing that gases are composed of molecules explains their obeying the Boyles-Charles law).

> Did our perceptions either inhere in something simple and individual, or did the mind perceive some real connexion among them, there wou'd be no difficulty in the case. For my part, I must plead the privilege of a sceptic, and confess, that this difficulty is too hard for my understanding. I pretend not, however, to pronounce it absolutely insuperable. Others, perhaps, or myself, upon more mature reflections, may discover some hypothesis, that will reconcile those contradictions.
>
> App.21; SBN 636

Of course, the hypothesis that would reconcile the contradictions that concern Hume would be one that would somehow explain the principles of connection among our perceptions that Hume discovers, but not employ a perceptible model for the ontology of the self in doing so.[17]

To summarize, then, it is impossible to explain the principle of connection among our perceptions because to do so would require positing a kind of substance for which we can have no perceptible model. That fact, however, does not eliminate the explanatory demand created by Hume's discovering such a principle,[18] and so Hume finds himself endorsing two contradictory theses. We must explain these principles because providing such explanations is at the very core of the scientific enterprise on which Hume embarks in the *Treatise*, and yet the results of that endeavor have ruled out precisely the kind of explanation that would be required.

With that, so closes my case for reading Hume as a kind of scientific realist. As I understand it, such a reading is demanded to make good philosophical sense of Hume's treatment of his most fundamental distinctions, is consistent with his explicit statements about his understanding of the science of human nature and the methodology that he in fact employs in pursuing this science, is also consistent with the most plausible interpretation of his theory of mental representation, and yields a valid reading of some of his most famous arguments. That interpretation also makes Hume's philosophical system as a whole more tenable and in line with what the philosophical understanding of science that was operative in Hume's own time and today. Thus, I conclude that Hume was a scientific realist who held that science explains the inductively-established universal generalizations discovered in experience by appeal to the nature, powers, and essence of the substance underlying these manifest phenomena.

Notes

1. The missing inconsistency was noticed early on in Kemp Smith, *The Philosophy of David Hume*, 558.
2. In 2010, Garrett reported that he was able to count no less than two dozen distinct accounts in the literature of what the inconsistency is supposed to be. Garrett, "Once More into the Labyrinth." Some such accounts include Robison, "Hume on Personal Identity"; Garrett, "Hume's Self-Doubts about Personal Identity"; Fogelin, *Hume's Skepticism*; Pears, *Hume's System*; Swain, "Being Sure of Oneself"; Loeb, "Causation, Extrinsic Relations, and Hume's Second Thoughts about Personal Identity"; Waxman, "Hume's Quandary Concerning Personal Identity"; Baxter, "Hume's Labyrinth Concerning the Idea of Personal Identity"; Roth, "What Was Hume's Problem with Personal Identity," and Kail, *Projection and Realism.*
3. Swain, "Being Sure of Oneself," also notes the importance of Hume's framing his concerns in terms of their explanatory adequacy, but develops this insight in a very different direction than I will. Swain leverages this point in an effort to show that Hume finds his account inadequate only according to the dubious explanatory standards set by the ancient and modern philosophies. I agree with Swain that it is crucial to see that what is at issue in the Appendix is specifically the explanatory adequacy of Hume's account, but I side with other commentators against Swain in holding that Hume finds something lacking according to the very standards that he himself sets.
4. Waxman, "Hume's Quandary Concerning Personal Identity," and Baxter, "Hume's Labyrinth Concerning the Idea of Personal Identity," likewise emphasize Hume's concern in the Appendix with the impossibility of forming an *idea* of the self, as opposed to a concern with the metaphysics of personal identity. Their accounts of representation are more restrictive than the one for which I

have argued, although our conclusions are similar: that it is the limits of representation as Hume conceives it that generate his concerns in the Appendix.

5. Strawson likewise emphasizes Hume's recognition of an ineliminable demand for explanation in the Appendix and notes that the *explicandum* there is the principle of connection (or the descriptive account that Hume offers in the body of the *Treatise*) itself. Strawson's take is that the problem that Hume confronts in the Appendix is that the principles of connection require a "substantive metaphysical characterization" that Hume's empiricism forbids. Strawson, *Evident Connexion*, 120–1. Strawson emphasizes that it is not the inability to *represent* such "real connections" that makes these out of bound for Hume—he takes it as obvious that Hume holds that we *can* represent such things despite evidence of the sort that I will present in the following section—but rather that it is a methodological commitment to his particular brand of Empiricism that does so. By way of explanation, he imagines an interlocutor raising the following accusation to Hume's former self.

> In relying on the I-Principles as you do you take a metaphysical step you can't take. You incur a certain general metaphysical debt you can't repay on your own empiricist principles. You can't rely on the I-Principles as you do and simply refer everything else to the unknown essence of the mind, for you can't stop someone replying that your reliance on the I-Principles entails that there is at least one thing that can be known about the essence of the mind and that you can't allow to be known.
>
> Strawson, *Evident Connexion*, 151

So, the demand for explanation with which Strawson takes Hume to struggle is really the demand to consistently carry through on his Empiricist commitments. That is, it is not, in fact, a demand to explain the principles of connection, but rather to cease and desist in appealing to them altogether because they are "real connections" of the kind that Hume more or less explicitly rejects. Thus, despite citing the idiom of a demand for explanation raised by these principles, Strawson's interpretation does not end up taking this demand very seriously at all. It is not as if Hume *could* make appeal to such metaphysically robust relations if only he could *explain* them. Rather, as Strawson sees it, Hume's commitments preclude him from such an appeal altogether. Hume might be able to explain such principles *away*, for example by reducing these to something more that carries the proper Empiricist *bona fides*, but such an explanation is an explanation in name alone. In this case, the demand would be less for an explanation, and more for an elimination.

6. Garrett, "Hume's Self-Doubts about Personal Identity," and Waxman, "Hume's Quandary Concerning Personal Identity," also locate Hume's concerns as centering on the notion of the unity of consciousness, although in very different ways. Garrett understands these as concerning the metaphysical unity of consciousness in light of the possibility of qualitatively identical perceptions belonging to numerically distinct minds. Waxman understands them as concerning the conditions that Hume's account of the representation of a bundle of perceptions itself requires for compresence of successive perceptions in a unified consciousness.
7. See also, 1.4.5.33; SBN 250: "To pronounce, then, the final decision upon the whole; the question concerning the substance of the soul is absolutely unintelligible."
8. We also followed Garrett in observing that the members of a Revival Set, and thereby the meaning of a definition, can be given in either of two ways: by

delineating either the *representing* ideas that form the Revival Set itself, or those perceptions that are *represented by* those ideas.

9. See 1.4.5.1; SBN 232.
10. Notice that this too parallels the case of the vulgar's idea of body. There it is the imagination's susceptibility to confirmation bias that leads it to suppose that our perceptions exhibit a degree of constancy and coherence that they do not. One can imagine a similar story here regarding the attention that proponents of this view of personal identity pay to those of our perceptions that appear to be constant and invariable.
11. One might object that Hume, in fact, limits his claim to the denial of the thesis that the self is *simple*, and so needs only the comparatively straightforward argument. Notice though that when Hume refers to the, "idea of *self*, after the manner it is here explain'd," i.e., as it is explained in the preceding paragraph, he is referencing, "its perfect *identity* and simplicity" (emphasis mine). The thesis that he will refute is not only that the self is simple, but also that it is complex, but numerically identical across time.
12. Of course, Hume also holds that these associative mechanisms are the only ones to be discovered in the human mind, and that as such, because of his acceptance of the Representational Copy Principle, they constitute the limits of our representational capacities.

> Every idea of a quality in an object passes thro' an impression; and therefore every *perceivable* relation, whether of connexion or repugnance, must be common both to objects and impressions.
>
> T 1.4.5.21; SBN 243.

> We cannot represent any relation among objects, much less between objects and subjects, that is not also a relation among perceptions.

13. Garrett, *Cognition and Commitment*, 49–50.
14. Strawson, *Evident Connexion*, parses Hume's idiom of "intelligibility" as indicating that we can give no coherent account of such an idea, not that we can have no such idea whatsoever. While Strawson might be able to apply this gloss more generally, Hume does also write in that same paragraph that we have *no idea of the self* as it is conceived by his predecessors, and that any answer to the question of the origin of such an idea would result in "manifest contradiction and absurdity".
15. Stroud, *Hume*, 124–6 presents several concerns regarding even the empirical regularities that Hume claims to find, and also the use to which he puts these in accounting for the idea of the self.
16. Thus, I agree with Stroud, Garrett, and Strawson that Hume's concern in the Appendix is not with what Strawson calls the Problem of Detail, "with the details of the psychological account of how we come to believe in a persisting self" (Strawson, *Evident Connexion*, 117). On the other hand, I also agree with Craig and Strawson (against Stroud) that Hume's concern in the Appendix is not with what Strawson calls the Bundling Problem, "how to get experiences into bundles, or how or why they come in bundles" (Strawson, *Evident Connexion*, 120). Stroud takes Hume's concern to be that the principles of connection cannot explain "the origin of the idea of the self" Stroud, *Hume*, 134. As Stroud sees it, this is because there is no way to specify what perceptions are surveyed in discovering these principles that does not non-circularly restrict the scope of this survey to oneself. Thus, Stroud does not take Hume at his word that his concern is with explaining the principles of connection *themselves*, but

rather relocates Hume's concern to one with the work that those principles can do in the service of explaining something else (the origin of our idea of the self). The philosophical issues that Stroud raises in this context are insightful and important ones, but the current account has the advantage over Stroud's (as it does over others) as an understanding of Hume's own concern that it is more faithful to the texts in which Hume expresses his concern.

17. It just so happens, of course, that this is precisely what Kant attempts to do in Transcendental Deduction and Paralogisms of the *Critique of Pure Reason*, wherein he argues that "the analytical unity of apperception is only possible under the presupposition of some synthetic one" (B131), and that the synthetic unity of apperception is "not even a representation distinguishing a particular object but rather a form of representation in general, insofar as it is to be called a cognition" (A346/B4040). See Landy, *Kant's Inferentialism*, Chapter 6.
18. In Chapter 2 I argued that Hume holds that in cases in which no theoretical representation is possible, one must resist the intemperate desire to offer illegitimate explanations, i.e., one must take certain facts to be explanatorily brute or primitive. Recalling that, one might suggest that rather than read Hume as confronted with a contradiction here between the demand for an explanation of the principles of connection of our perceptions and the impossibility of giving such an explanation, one might resolve this tension by simply taking these principles of connection to be primitive in the way suggested in Chapter 2. What this brings out is that on the interpretive line that I have been pursuing, it is only non-universal regularities that can be taken as primitive in this way, and this is precisely because they do not really demand explanation at all. The examples discussed earlier fit this pattern. For example, the principles of association are not universal and are taken as explanatorily basic; the apparent constancy and coherence of our perceptions are not universal and demand no explanation, etc.

Bibliography

Allison, Henry. *Custom and Reason in Hume: A Kantian Reading of the First Book of the Treatise*. Oxford: Oxford University Press, 2008.

Baxter, Donald. "Abstraction, Inseparability, and Identity." *Philosophy and Phenomenological Research* 57 (1997): 307–30.

Baxter, Donald. "Hume's Labyrinth Concerning the Idea of Personal Identity." *Hume Studies* 24 2 (1998): 203–34.

Beebee, Helen. *Hume on Causation*. New York: Routledge, 2006.

Bennett, Jonathan. *Learning from Six Philosophers, Vol. 2*. Oxford: Oxford University Press, 2001.

Berkeley, George. *The Works of George Berkeley, Vol. 1 of 4*. Oxford: Clarendon Press, 1901.

Blackburn, Simon. "Hume and Thick Connections." *Philosophy and Phenomenological Research* 1 (1990): 237–50.

Boehm, Miren. "The Concept of *Body* in Hume's *Treatise*." *ProtoSociology* 30 (2013): 206–20.

Boehm, Miren. "Hume's Foundational Project in the *Treatise*." *European Journal of Philosophy* 24 (2016): 55–77.

Broughton, Janet. "Hume's Ideas about Necessary Connection." *Hume Studies* 13 2 (1987): 217–44.

Butler, Annemarie. "Hume on Believing the Vulgar Fiction of Continued Existence." *History of Philosophy Quarterly* 27 3 (2010): 237–54.

Cohon, Rachel and Owen, David. "Hume on Representation, Reason, and Motivation." *Manuscrito* 20 2 (1997): 47–76.

Craig, E. J. *The Mind of God and the Works of Man*. Oxford: Clarendon Press, 1987.

Cummins, Phillip D. "Hume on the Idea of Existence." *Hume Studies* 17 1 (1991): 61–82.

De Pierris, Graciela. "Causation as a Philosophical Relation in Hume." *Philosophy and Phenomenological Research* 64 (2002): 499–545.

De Pierris, Graciela. "Hume and Locke on Scientific Methodology: The Newtonian Legacy." *Hume Studies* 32 (2006): 277–330.

De Pierris, Graciela. *Ideas, Evidence, and Method: Hume's Skepticism and Naturalism Concerning Knowledge and Causation*. Oxford: Oxford University Press, 2015.

Flage, Daniel. "Hume on Memory and Causation." *Hume Studies* 11 Supplement: 10th Anniversary Issue (1985): 168–88.

Flage, Daniel. "Hume's Relative Ideas." *Hume Studies* 7 1 (1981): 55–73.

Flage, Daniel. "Perchance to Dream: A Reply to Traiger." *Hume Studies* 11 2 (1985): 173–82.

Flage, Daniel. "Relative Ideas Re-Viewed." In *The New Hume Debate*. Rev. ed. Edited by Rupert Read and Kenneth A. Richman, 138–55. New York: Routledge, 2000.
Fogelin, Robert. *Hume's Skepticism in the Treatise of Human Nature*. London: Routledge & Kegan Paul Ltd., 1985.
Friedman, Lesley. "Another Look at Flage's Hume." *Hume Studies* 19 2 (1993): 177–86.
Gamboa, Steven. "Hume on Resemblance, Relevance and Representation." *Hume Studies* 33 1 (2007): 21–40.
Garrett, Don. *Cognition and Commitment in Hume's Philosophy*. Oxford: Oxford University Press, 1997.
Garrett, Don. "Hume's Naturalistic Theory of Representation." *Synthese* 152 3 (2006): 301–19.
Garrett, Don. "Hume's Self-Doubts about Personal Identity." *The Philosophical Review* 90 3 (1981): 337–58.
Garrett, Don. "Once More into the Labyrinth: Kail's Realist Explanation of Hume's Second Thoughts about Personal Identity." *Hume Studies* 36 1 (2010): 77–88.
Garrett, Don. "'A Small Tincture of Pyrrhomism' Skepticism and Naturalism in Hume's Science of Man." In *Pyrrhonian Skepticism*. Edited by Walter Sinnott-Armstrong, 68–98. Oxford: Oxford University Press, 2004.
Hawkins, R. J. "Simplicity, Resemblance and Contrariety in Hume's *Treatise*." *Philosophical Quarterly* 26 (1976): 24–38.
Hazony, Yoram. "Newtonian Explanatory Reduction and Hume's System of the Sciences." In *Newton and Empiricism*. Edited by Zvi Biener and Eric Schliesser, 138–70. Oxford: Oxford University Press, 2014.
Hazony, Yoram and Schliesser, Eric. "Newton and Hume." In *The Oxford Handbook of Hume*. Edited by Paul Russell, 673–707. Oxford: Oxford University Press, 2016.
Hume, David. *Dialogues Concerning Natural Religion*. Edited by Norman Kemp Smith. Edinburgh: Nelson, 1947.
Hume, David. *A Dissertation on the Passions* and *The Natural History of Religion*. Edited by Tom L. Beauchamp. Oxford: Clarendon Press, 2007.
Hume, David. *Enquiries Concerning the Human Understanding and Concerning the Principles of Morals*. Edited by L. A. Selby-Bigge, Revised by P. H. Nidditch. Oxford: Clarendon Press, 1975.
Hume, David. *An Enquiry Concerning Human Understanding: A Critical Edition*. Edited by Tom L. Beauchamp. Oxford: Clarendon Press, 2001.
Hume, David. *Essays: Moral, Political, and Literary*. Edited by Eugene F. Miller. Liberty Fund, Inc., 1987.
Hume, David. *Essays on Suicide and the Immortality of the Soul*. Bristol: Thoemmes Press, 1992.
Hume, David. *A Treatise of Human Nature*. Edited by L. A. Selby-Bigge. New York: Oxford University Press, 1974.
Hume, David. *A Treatise of Human Nature*. Edited by David Fate Norton and Mary J. Norton. New York: Oxford University Press, 2000.
Kail, Peter. *Projection and Realism in Hume's Philosophy*. New York: Oxford University Press, 2010.
Kant, Immanuel. *Critique of Pure Reason*. Translated by Paul Guyer and Alan Wood. Cambridge: Cambridge University Press, 2008.
Kemp Smith, Norman. *The Philosophy of David Hume*. New York: Macmillan, 1941.
Kripke, Saul. "Speaker's Reference and Semantic Reference." *Midwest Studies in Philosophy* 2 (1977): 255–76.

Landy, David. "Hume's Impression-Idea Distinction." *Hume Studies* 32 (2006): 119–39.

Landy, David. "Hume's Theory of Mental Representation." *Hume Studies* 38 1 (2012): 23–54.

Landy, David. *Kant's Inferentialism: The Case against Hume*. New York: Routledge, 2015.

Landy, David. "A Puzzle about Hume's Theory of General Representation." *Journal of the History of Philosophy* 54 2 (2016): 257–82.

Landy, David. "Recent Scholarship on Hume's Theory of Mental Representation." *European Journal of Philosophy* (forthcoming).

Loeb, Louis E. "Causation, Extrinsic Relations, and Hume's Second Thoughts about Personal Identity." *Hume Studies* 18 2 (1992): 219–32.

Loeb, Louis E. *Stability and Justification in Hume's Treatise*. Oxford: Oxford University Press, 2002.

Mackie, J. L. "Locke's Anticipation of Kripke." *Analysis* 34 6 (1974): 188–80.

McDonough, Jeffrey K. "Hume's Account of Memory." *British Journal for the History of Philosophy* 10 1 (2002): 71–87.

Millican, Peter. "Hume, Causal Realism, and Causal Science." *Mind* 118 (2009): 647–712.

Millikan, Ruth Garrett. *Language, Thought, and Other Biological Categories*. Cambridge, MA: The MIT Press, 1987.

Nelson, Alan and Landy, David. "Qualities and Simple Ideas: Hume and His Debt to Berkeley." In *Primary and Secondary Qualities: The Historical and Ongoing Debate*. Edited by Lawrence Nolan, 216–38. Oxford: Oxford University Press, 2011.

Noonan, Harold W. *Routledge Philosophy Guidebook to Hume on Knowledge*. New York: Routledge, 1999.

Owen, David. *Hume's Reason*. Oxford: Oxford University Press, 1999.

Pears, David. *Hume's System*. Oxford: Oxford University Press, 1991.

Qu, Hsueh. "Hume on Mental Transparency." *Pacific Philosophical Quarterly* (forthcoming).

Qu, Hsueh. "Hume's (Ad Hoc?) Appeal to the Calm Passions." *Archiv für Geschichte der Philosophie* (forthcoming).

Qu, Hsueh. "Prescription, Description, and Hume's Experimental Method." *The British Journal for the History of Philosophy* 24 2 (2016): 279–301.

Robison, Wade. "Hume on Personal Identity." *Journal of the History of Philosophy* 12 2 (1974): 181–93.

Rosenberg, Jay F. *Beyond Formalism: Naming and Necessity for Human Beings*. Philadelphia: Temple University Press, 1994.

Roth, Abe. "What Was Hume's Problem with Personal Identity." *Philosophy and Phenomenological Research* 61 1 (2000): 91–114.

Russow, Lilly-Marlene. "Simple Ideas and Resemblance." *The Philosophical Quarterly* 30 (1980): 342–50.

Salmon, Wesley. "Four Decades of Scientific Explanation." In *Scientific Explanation*. Edited by Philip Kitcher and Wesley C. Salmon, 3–219. Minneapolis: University of Minnesota Press, 1989.

Schliesser, Eric. "Hume's Attack on Newton's Philosophy." *Enlightenment & Dissent* 25 (2009): 167–203.

Schliesser, Eric. "Hume's Newtonianism and Anti-Newtonianism." In *The Stanford Encyclopedia of Philosophy* (Winter 2008 Edition). Edited by Edward N. Zalta. http://plato.stanford.edu/archives/win2008/entries/hume-newton/.

Sellars, Wilfrid. "Empiricism and the Philosophy of Mind." In *Science, Perception and Reality*. 119–79. Atascadero, CA: Ridgeview Publishing Co., 1963.

Sellars, Wilfrid. "The Language of Theories." In *Science, Perception and Reality*. 100–18. Atascadero, CA: Ridgeview Publishing Co., 1963.

Sellars, Wilfrid. "Naming and Saying." In *Science, Perception and Reality*. 207–27. Atascadero, CA: Ridgeview Publishing Co., 1963.

Sellars, Wilfrid. *Science and Metaphysics*. Atascadero, CA: Ridgeview Publishing Co., 1967.

Sellars, Wilfrid. "Scientific Realism or Irenic Instrumentalism: A Critique of Nagel and Feyerabend on Theoretical Explanation." *Boston Studies in the Philosophy of Science* 2 (1965): 171–201.

Strawson, Galen. *The Evident Connexion*. Oxford: Oxford University Press, 2011.

Strawson, Galen. *The Secret Connexion*. Rev. ed. Oxford: Oxford University Press, 2014.

Stroud, Barry. *Hume*. London: Routledge & Kegan Paul Ltd., 1977.

Swain, Corliss. "Being Sure of Oneself: Hume on Personal Identity." *Hume Studies* 17 2 (1991): 107–24.

Tienseon, John. "Resemblance and General Terms." *Philosophical Studies* 54 (1988): 87–108.

Traiger, Saul. "Flage on Hume's Account of Memory." *Hume Studies* 11 2 (1985): 166–72.

van Fraassen, Bas. *The Scientific Image*. Oxford: Oxford University Press, 1980.

Vision, Gerald. "Hume's Attack on Abstract Ideas: Real and Imagined." *Dialogue* 18 (1979): 528–37.

Waxman, Wayne. "Hume's Quandary Concerning Personal Identity." *Hume Studies* 18 2 (1992): 233–54.

Weinberg, Julius R. *An Examination of Logical Positivism*. Abingdon: Routledge & Kegan Paul Ltd., 1936.

Wilson, Fred. "Is Hume a Sceptic with Regard to the Senses?" *Journal of the History of Philosophy* 27 1 (1989): 49–73.

Winkler, Kenneth. "The New Hume." *The Philosophical Review* 100 (1991): 541–79.

Wittgenstein, Ludwig. *Tractatus Logico-Philosophicus*. Translated by C. K. Ogden. London: Kegan Paul, Trench, Trubner & Co. Ltd., 1922.

Wolff, Robert Paul. "Hume's Theory of Mental Activity." *The Philosophical Review* 69 (1960): 289–310.

Wright, John P. *The Sceptical Realism of David Hume*. Manchester: Manchester University Press, 1983.

Index

For Product Safety Concerns and Information please contact our EU representative GPSR@taylorandfrancis.com
Taylor & Francis Verlag GmbH, Kaufingerstraße 24, 80331 München, Germany

www.ingramcontent.com/pod-product-compliance
Ingram Content Group UK Ltd.
Pitfield, Milton Keynes, MK11 3LW, UK
UKHW022002180926
475701UK00003B/150
9780367891718